Undergraduate Texts in Mathematics

Editors
S. Axler
F.W. Gehring
K.A. Ribet

T0155695

Springer
New York
Berlin
Heidelberg
Barcelona
Hong Kong
London
Milan
Paris
Singapore
Tokyo

Undergraduate Texts in Mathematics

Abbott: Understanding Analysis.
Anglin: Mathematics: A Concise History and Philosophy.
Readings in Mathematics.
Anglin/Lambek: The Heritage of Thales.
Readings in Mathematics.
Apostol: Introduction to Analytic Number Theory. Second edition.
Armstrong: Basic Topology.
Armstrong: Groups and Symmetry.
Axler: Linear Algebra Done Right. Second edition.
Beardon: Limits: A New Approach to Real Analysis.
Bak/Newman: Complex Analysis. Second edition.
Banchoff/Wermer: Linear Algebra Through Geometry. Second edition.
Berberian: A First Course in Real Analysis.
Bix: Conics and Cubics: A Concrete Introduction to Algebraic Curves.
Brémaud: An Introduction to Probabilistic Modeling.
Bressoud: Factorization and Primality Testing.
Bressoud: Second Year Calculus.
Readings in Mathematics.
Brickman: Mathematical Introduction to Linear Programming and Game Theory.
Browder: Mathematical Analysis: An Introduction.
Buchmann: Introduction to Cryptography.
Buskes/van Rooij: Topological Spaces: From Distance to Neighborhood.
Callahan: The Geometry of Spacetime: An Introduction to Special and General Relavitity.
Carter/van Brunt: The Lebesgue–Stieltjes Integral: A Practical Introduction.
Cederberg: A Course in Modern Geometries. Second edition.

Childs: A Concrete Introduction to Higher Algebra. Second edition.
Chung: Elementary Probability Theory with Stochastic Processes. Third edition.
Cox/Little/O'Shea: Ideals, Varieties, and Algorithms. Second edition.
Croom: Basic Concepts of Algebraic Topology.
Curtis: Linear Algebra: An Introductory Approach. Fourth edition.
Devlin: The Joy of Sets: Fundamentals of Contemporary Set Theory. Second edition.
Dixmier: General Topology.
Driver: Why Math?
Ebbinghaus/Flum/Thomas: Mathematical Logic. Second edition.
Edgar: Measure, Topology, and Fractal Geometry.
Elaydi: An Introduction to Difference Equations. Second edition.
Exner: An Accompaniment to Higher Mathematics.
Exner: Inside Calculus.
Fine/Rosenberger: The Fundamental Theory of Algebra.
Fischer: Intermediate Real Analysis.
Flanigan/Kazdan: Calculus Two: Linear and Nonlinear Functions. Second edition.
Fleming: Functions of Several Variables. Second edition.
Foulds: Combinatorial Optimization for Undergraduates.
Foulds: Optimization Techniques: An Introduction.
Franklin: Methods of Mathematical Economics.
Frazier: An Introduction to Wavelets Through Linear Algebra.
Gamelin: Complex Analysis.
Gordon: Discrete Probability.
Hairer/Wanner: Analysis by Its History.
Readings in Mathematics.
Halmos: Finite-Dimensional Vector Spaces. Second edition.

(continued after index)

George E. Martin

Counting: The Art of Enumerative Combinatorics

 Springer

George E. Martin
Department of Mathematics and Statistics
State University of New York at Albany
Albany, NY 12222
USA
martin@math.albany.edu

Mathematics Subject Classification (2000): 05-01, 05Axx

Library of Congress Cataloging-in-Publication Data
Martin, George Edward, 1932–
    Counting : the art of enumerative combinatorics / George E. Martin.
      p. cm. —- (Undergraduate texts in mathematics)
    Includes bibliographical references and index.
    ISBN 978-1-4419-2915-0
    1. Combinatorial enumeration problems. I. Title. II. Series.

    QA164.8 .M37 2001
    511'.62—dc21   00-067918

Printed on acid-free paper.

Production managed by Steven Pisano; manufacturing supervised by Jeffrey Taub.
Photocomposed pages prepared from the author's LaTeX files.

Printed in the United States of America.

9 8 7 6 5 4 3 2 1

Springer-Verlag   New York Berlin Heidelberg
A member of BertelsmannSpringer Science+Business Media GmbH

To Margaret

Books by the Author

The Foundations of Geometry
and the Non-Euclidean Plane

Transformation Geometry:
An Introduction to Symmetry

Polyominoes:
A Guide to Puzzles and Problems in Tiling

Geometric Constructions

Counting:
The Art of Enumerative Combinatorics

Preface

Among the titles of the several class-tested versions of this book were the alliterative sounding *Counting: A College Course in Combinatorics* and the descriptive *Counting as a First Course in Discrete Mathematics: An Introduction to Enumerative Combinatorics*, which sounded like something from the nineteenth century. These titles and the final selection explain the intent of the text. The details follow. Chapters 1 and 2 cover the elements, including the principle of inclusion and exclusion. Chapter 3 deals with generating functions. These three chapters form the core. The dependency of the chapters is as follows.

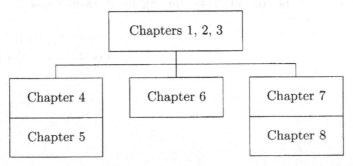

Chapters 4 (Groups) and 5 (Actions) take us through the Pólya pattern inventory theory. The chapter on groups introduces symmetry groups, permutation groups, and what is necessary and sufficient for the practical applications of group theory used in the next chapter. (I have not noticed that those who have already had abstract algebra have a distinct advantage

over those who have not. On the contrary, I have been told that this material has helped those students who have gone on to take abstract algebra. If there is a small overlap with other courses, it appears to be a beneficial overlap.) Applications of Burnside's lemma and the Pólya theory constitute Chapter 5. This is a nontrivial piece of mathematics that students seem to enjoy thoroughly. The topic of Chapter 6 is recurrence relations. Chapter 7 deals with mathematical induction, and Chapter 8 is a brief introduction to graph theory.

I have used preliminary versions of this text for a semester course at the sophomore level called An Introduction to Discrete Mathematics for the Department of Computer Science and for the Department of Mathematics and Statistics. This course usually concentrated on Chapters 1, 2, 3, and 6, with bits of 7 and 8 thrown in. I have used the first five chapters as a graduate course for in-service teachers under the rubric Algebra for Teachers. Altogether, there is enough material for a year course.

Beginning enumerative combinatorics necessarily requires a problem-solving approach. In the beginning, there is little theory but a lot of mathematical maturity to be learned. An instructor should not be upset that there are very few problems that do not have answers given in the part that is called The Back of the Book. Both student and instructor will find this important part of the book very useful. I have found that fair but tough exams can be constructed on exactly the same questions that are assigned for homework.

What is a geometer doing writing a book on combinatorics? Actually, I started out doing mathematical research in the area of finite projective planes, which is combinatorics in geometric language. Such planes are sets of mutually orthogonal latin squares in disguise. Although I next turned to foundations of geometry, in recent years I have come to tiling the plane as my principle interest. However, in addition to geometry courses, I have enjoyed teaching the content of this text for many, many years.

George E. Martin
martin@math.albany.edu

Contents

Chapter 3. Generating Functions

Chapter 4. Groups

Chapter 5. Actions

Chapter 6. Recurrence Relations

Chapter 7. Mathematical Induction

Chapter 8. Graphs

1
Elementary Enumeration

§1. Counting Is Hard

Yes, counting *is* hard. We may as well get that out and understood right at the top. "Counting" is short for "enumerative combinatorics," which certainly doesn't sound easy. This is a course in discrete mathematics that addresses questions that begin, *How many ways are there to* ... . For example, we shall soon know the answer to questions such as, How many ways are there to order 12 ice cream cones if 8 flavors are available? At the end of the course we should be able to answer such nontrivial counting questions as, How many ways are there to color the faces of a cube if k colors are available, with each face having exactly 1 color? or How many ways are there to stack n poker chips, each of which can be red, white, blue, or green, such that each red chip is adjacent to at least 1 green chip?

There are no prerequisites for this course beyond mathematical maturity. Of course, one gets mathematical maturity by taking mathematics courses. This is as good a place to start as any other part of mathematics. One of the things that make elementary counting difficult is that we will encounter very few algorithms. You will have to think. There are few formulas and each problem seems to be different. Unfortunately, the only way to learn to do elementary counting problems is to do a lot of elementary counting problems. Fortunately, this pays off for most people. Mastering the ways of looking at a counting problem comes only with practice, practice, and more practice.

Also, counting can be fun.

§2.   Conventions

In order to shorten descriptions as much as possible, we will adopt some conventions that will hold throughout the book. Persons are always distinguishable, one from another. For our purposes, oranges are indistinguishable, one from another. The same can be said for apples, A's, red balls, and, unless otherwise specified, even dimes. You are encouraged to generalize and apply the conventions to peaches, D's, and green balls, for example. Order within a choice is not to be considered unless specifically mentioned. There are 26 letters in the English alphabet; we ignore such words as "rhythm" and "cwm" and boldly declare that there are 5 vowels (a, e, i, o, u) and 21 consonants. The word "word" is used much as it is in computer science; here, a word is just a string of symbols, which, unless otherwise specified, are assumed to be letters of the alphabet. We certainly do not expect to find most such words in an English dictionary. Coins are American coins and come in the denominations: penny (1¢), nickel (5¢), dime (10¢), quarter (25¢), half dollar (50¢), and dollar ($1 = 100¢).

For nonnegative integer n, the symbol $n!$ is not shouted but read "n factorial." Factorials are defined recursively by $0! = 1$ and $n! = [n][(n-1)!]$ for positive integers n. So, for example, $5! = [5][4!] = 5 \times 4 \times 3 \times 2 \times 1$.

A deck of cards consists of 52 cards, 13 in each of 4 suits: spades ♠, clubs ♣, hearts ♡, and diamonds ◇. Spades and clubs are black; hearts and diamonds are red. Each suit has a card of each of the face values 2, 3, 4, ... , 10, jack, queen, king, ace. Cards of the same face value are said to be of the same kind. A bridge hand contains 13 cards, while a poker hand contains 5 cards. A die is a cube with the sides numbered from 1 to 6.

How many months have 28 days? The answer to this old riddle is, of course, All of them. Usually, in mathematics and logic, to say that there are 10 balls in the box means that there are *at least* 10 balls in the box. If there are 12 balls in the box, then there are certainly 10 balls in the box. However, in combinatorics and probability, as well as "on the street," to say that there are 10 balls in the box means that there are *exactly* 10 balls in the box. When we talk about distributing 10 balls into 5 boxes, we mean exactly 10 balls are to be put into the 5 boxes. This prompts the admission that we realize that most of us will not spend large amounts of our life putting balls into boxes, although we spend a lot of time in this text talking about putting balls into boxes. We are happy to abstract placement problems to balls and boxes. This does not diminish the seriousness of the mathematical applications.

Ten Quickies.

Answer the following 10 questions. Then check in The Back of the Book, which begins on page 183, to see how well you have done.

1. How many ways are there to pick 1 student from 6 boys and 8 girls?

2. How many ways are there to pick 1 piece of fruit from 6 oranges and 8 apples?

3. How many ways are there to pick 1 letter from 3 A's, 5 B's, and 7 C's?

4. How many ways are there to pick 2 letters from 3 B's and 3 G's?

5. How many ways are there to pick 2 students from 3 boys and 3 girls?

6. How many ways are there to pick 5 oranges from 6 oranges?

7. How many ways are there to pick 5 girls from 6 girls?

8. How many ways are there to pick 1 girl from 6 girls?

9. How many ways are there to pick 5 pieces of fruit from 7 oranges and 8 apples?

10. How many ways are there to pick some pieces of fruit from 9 oranges and 6 apples if at least 1 piece is picked?

§3. Permutations

More Quickies.

1. How many ways are there to pick a Latin book and a Greek book from 5 distinguishable Latin books and 7 distinguishable Greek books?

2. How many ways are there to make a 2-letter word?

3. How many ways are there to make a 2-letter word if the letters must be different?

4. How many ways are there to make a 2-letter word with a consonant followed by a vowel?

5. How many ways are there to pick a boy and a girl from 3 boys and 8 girls?

6. How many ways are there for 2 persons to sit in 5 chairs that are in a row?

7. How many ways are there to pick 2 of 5 chairs that are in a row?

8. How many ways are there to make a 4-letter word?

9. How many ways are there to pick an element from the 5-by-7 matrix (a_{ij})?

10. How many ways are there to pick an element from the m-by-n matrix (a_{ij})?

Observation. *The Multiplication Principle:* If one thing can be done in m ways and a second thing can be done in n ways independent of how the first thing is done, then the 2 things can be done in mn ways.

11. How many ways are there to flip a coin and toss a die?

12. How many ways are there to flip a coin, toss a die, and pick a card from a deck of cards?

13. How many ways are there to arrange the aces from a deck of cards in a row?

14. How many ways are there to arrange the spades from a deck of cards in a row?

15. How many ways are there to arrange (in a row) all n elements of the set $\{a_1, a_2, a_3, \dots, a_n\}$? Each of these arrangements is called a **permutation** of the elements in the set. An arrangement using exactly r of the n elements is called an r-**permutation** of the elements in the set. It follows that there are

$$(n - 0)(n - 1)(n - 2) \cdots (n - (r - 1))$$

r-permutations of n distinguishable objects.

Observation. The number of permutations of n distinguishable objects is $n!$. The number of r-permutations of n distinguishable objects is $n!/(n-r)!$.

§4.   A Discussion Question

How many ways can a pair of dice fall?

§5.   The Pigeonhole Principle

If there are more pigeons than pigeonholes, then some pigeonhole must contain at least 2 pigeons. To be more specific, if $n + 1$ or more pigeons are assigned to n pigeonholes, then at least 2 pigeons are assigned to the same pigeonhole. To be more general, if there are more than k times as many pigeons as pigeonholes, then some pigeonhole must contain at least $k + 1$ pigeons. This doesn't seem like advanced mathematics at first. Actually, most generalizations of the Pigeonhole Principle are beyond our goal of studying elementary combinatorics. We will follow an example by some pigeonhole problems that are fun but not exactly trivial.

For a nontrivial example, consider 6 points in the plane with no 3 collinear. Suppose each pair of these points is joined by either a red segment or a green segment. Show that at least 3 of the points are joined by segments of the same color. You may wish to put the book down and try this rather hard problem before reading about the solution. Now, at least 3 of the 5 segments joining point A in the set to the other 5 points must be of the same color by the Pigeonhole Principle. We call these point D, E, and F, and, without loss of generality, we may suppose the color is red. If 1 of the segments $\overline{DE}$, $\overline{EF}$, or $\overline{FD}$ is also red, then the problem is solved; and, on the other hand, if all 3 of these segments are green, the problem is also solved.

Pigeonhole Problems.

1. We have 10 indistinguishable white socks and 10 indistinguishable black socks in our sock drawer. How many socks must we withdraw from our sock drawer, without seeing the socks, in order to be sure that we have a matching pair?

2. How many cards must we draw from a deck of cards to be sure of getting at least 2 from the same suit?

3. How many people must be in a room in order to be sure that at least 3 have the same birthday?

4. How many balls must be chosen from 12 red balls, 20 white balls, 7 blue balls, and 8 green balls to be assured that there are 10 balls of the same color?

5. How many persons must be chosen from *n* couples in order to be sure that 1 couple is included?

6. Show that in a room of 20 persons there are at least 2 persons who have the same number of mutual friends in the room.

7. Consider the points in the plane that have integer coordinates. Show that at least 1 of the 10 segments joining any 5 of these points contains another such point.

8. Show that at any party of at least 6 persons there is a set of 3 that are mutual acquaintances or a set of 3 that are mutual strangers.

§6.   *n* Choose *r* by Way of MISSISSIPPI

As we will frequently do, we state a sequence of problems, each of which leads to the next. We may not have been able to answer the last question in the first place, but the answer to the last question becomes "obvious" after going through the other questions. In the following we assume that A_1, A_2, and A_3 are letters. It might help to think of them first as 3 A's having 3 different colors and later as indistinguishable A's. Likewise, first think of E_4 and E_5 as different colored E's.

1. How many ways are there to arrange the 6 letters of the word ABCDEF ?

2. How many ways are there to arrange the 6 letters of the word $A_1A_2A_3E_4E_5F$?

3. How many ways are there to arrange the 6 letters of the word $A_1A_2A_3EEF$?

4. How many ways are there to arrange the 6 letters of the word $AAAE_4E_5F$?

5. How many ways are there to arrange the 6 letters of the word AAAEEF ?

6. How many ways are there to arrange the letters of the word BANANA ?

7. How many ways are there to arrange the letters of the word AAABBCCCCD ?

8. How many ways are there to arrange the letters of the word MATHEMATICS

9. How many ways are there to arrange the letters of the word MISSISSIPPI ?

In doing this problem, the first thing to do is make the figure

$$M \ I \ S \ P$$
$$I \ S \ P$$
$$I \ S$$
$$I \ S$$

formed by spelling out the word letter by letter, starting a new column with each new letter encountered. (You will use this very figure many times in the sequel.) We may want to add an "11" to the bottom right to keep track of the total number of letters. Seeing that there are $1, 4, 4, 2$ letters in the respective columns, we calculate the answer $11!/[1!4!4!2!]$ to our question, where the "1" in the denominator is completely optional.

Problems like #5, 6, 7, 8, and 9 will be called **Mississippi problems.** Such problems are frequently part of a larger problem. We suppose now that we are quite capable of answering any Mississippi problem, regardless of the word given.

10. How many ways are there to arrange 4 A's, 3 G's, and the 6 letters U, V, W, X, Y, and Z?

11. How many ways are there to arrange on a shelf 4 copies of an algebra book, 3 copies of a geometry book, and 6 different trashy novels?

12. How many ways are there to arrange on a shelf 4 different algebra books, 3 different geometry books, and 6 different calculus books such that the books on each subject are grouped together?

13. How many n-letter words have r C's and $n - r$ R's?

Apply the solution of #13 to #14 where C stands for "choose" and R stands for "reject" for the individuals considered 1 at a time and standing in a row.

14. How many ways can we select r persons from n persons when $n \geq r$?

15. How many ways can we select r distinguishable objects from n distinguishable objects when $n \geq r$?

Each of the selections in #15 is called an r-**combination** of the n elements. An r-permutation is ordered, and an r-combination is unordered. Note that each r-combination corresponds to exactly $r!$ of the r-permutations. So, if we let $\binom{n}{r}$ denote the number of r-combinations of n objects, then $\binom{n}{r}r!$ must be $n!/(n-r)!$, the number of r-permutations of n objects. Sometimes $\binom{n}{r}$ is read as "(the number of) combinations of n things taken r at a time" and sometimes as "n above r." Note that "n over r" denotes n/r and is different from n above r. In any case, we recommend always reading $\binom{n}{r}$ as "n choose r."

Observation. The number of r-combinations of n objects is $n!/r!(n-r)!$. That is,

$$\binom{n}{r} = \frac{n!}{r!(n-r)!}. \quad \text{Clearly,} \quad \binom{n}{r} = \binom{n}{n-r}.$$

Can you think of a story that proves the following relationship for $0 < r < n$?

$$\binom{n}{r} = \binom{n-1}{r-1} + \binom{n-1}{r}.$$

How about the following story, which also introduces our very handy friend Lucky Pierre? We are to select a committee of r persons from a group of n persons, 1 of whom is Lucky Pierre. Now, there are $\binom{n-1}{r-1}$ ways to choose the committee with Lucky Pierre on the committee, and there are $\binom{n-1}{r}$ ways to choose the committee with Lucky Pierre not on the committee. The sum of these 2 numbers must be $\binom{n}{r}$, the number of ways of selecting the committee. (Of course, we can also prove the relation by expanding and simplifying both sides.)

Use the equation displayed above to continue, for at least 1 more line, the array that is called **Pascal's Triangle** and is illustrated in Table 1.1. Except for the 1's in Pascal's Triangle, each entry is the sum of the 2 integers that are to the right and left of it in the line above it.

§7.   The Round Table

Our convention for counting the number of seatings at a round table is that seatings s_1 and s_2 are considered the same seating iff (if and only if) everyone at the table has the same right-hand neighbor in s_1 as in s_2. That

```
                                        1
                                    1       1
                                1       2       1
                            1       3       3       1
                        1       4       6       4       1
                    1       5      10      10       5       1
                1       6      15      20      15       6       1
            1       7      21      35      35      21       7       1
        1       8      28      56      70      56      28       8       1
    1       9      36      84     126     126      84      36       9       1
1      10      45     120     210     252     210     120      45      10       1
1   11      55     165     330     462     462     330     165      55      11       1
1   12      66     220     495     792     924     792     495     220      66      12       1
```

TABLE 1.1. Pascal's Triangle

is, if each person at a round table moves to the seat at their right, then we count this as the same seating arrangement. All rotations of a particular arrangement are considered the same. The chairs are evenly distributed, and no one takes a chair until all persons have been placed around the table.

1. How many ways can 8 persons be seated at a round table?

It is quite handy to consider Lucky Pierre as 1 of the persons that is to be seated and place him first. Since this is a round table, it makes no difference which chair he sits in. So put Lucky Pierre at the table. Now, there is an order established for the remaining places, say, to Lucky Pierre's right. The remaining 7 persons can be seated then in 7! ways. The answer is 7!.

2. How many ways can 12 of King Arthur's knights be seated at a round table?

The answer is 11!. (Who knew that Lancelot was really Lucky Lancelot Pierre?) We can easily generalize to the following observation.

Observation. The number of ways of seating n persons at a round table is $(n-1)!$.

3. How many ways can 8 couples be seated in a row if each couple is seated together?

The persons in each couple can be arranged in 2! ways. Then the couples can be arranged in a row in 8! ways. The answer is $2^8 \times 8!$.

4. How many ways can 8 couples be seated at a round table if each couple is seated together?

The persons in each couple can be arranged in 2! ways. Then the couples, including, of course, the Pierres, can be arranged at the round table in 7! ways. The answer is $2^8 \times 7!$.

It is a considerable help in doing most counting problems to mentally perform the physical actions that are necessary to accomplish some desired result. In thinking about #4 above, for example, our good friends Lucy and Lucky Pierre are a convenient construct. We ask ourself, in placing the Pierres at the round table, Do I seat Lucky to the right or to the left of Lucy? Likewise for each of the other couples, Which member sits on the right of the other? Now, with the Pierres already at the round table, we ask, In what order can we place the remaining ordered couples to the right of the Pierres? Answering these easy questions, we get to the final answer. On the other hand, if we think of putting 1 place card down for each couple first and then permuting the 2 members of each couple, we get the answer $7! \times 2!^8$.

5. How many ways can 4 Americans, 7 Belgians, and 10 Canadians be seated at a round table?

Is the answer 20! obvious?

6. How many ways are there to arrange 4 C's and 8 R's such that no 2 C's are adjacent?

We first put down the 8 R's in a row. There is only 1 way to do this. Then, from the diagram $_\wedge R_\wedge R_\wedge R_\wedge R_\wedge R_\wedge R_\wedge R_\wedge R_\wedge$, we see that there are 9 spaces into each of which we can insert at most 1 of the 4 C's. We can choose these 4 spaces in $\binom{9}{4}$ ways and, of course, insert the C's into these spaces in 1 way. Thus, there are $\binom{9}{4}$ words with 4 C's and 8 R's such that no 2 C's are adjacent. We will find that the wedge $\wedge$ is an invaluable tool to be used to indicate places for spaces in many counting problems.

Homework.

1. How many ways are there to select a committee of 5 from 11 teachers?

2. How many poker hands (5 cards) are there?

3. How many bridge hands (13 cards) are there?

4. How many full houses (three-of-a-kind and a pair) are there in poker?

5. How many ways are there for John to invite some of his 10 friends to dinner, if at least 1 of the friends is invited?

6. How many ways are there to arrange the letters of the word DABBADABBADO?

7. There are 5 algebra books, 7 geometry books, and 4 calculus books. The books are distinguishable. How many ways are there to pick 2 books not both on the same subject?

8. How many different selections can be made from 5 apples and 7 oranges if at least 1 piece of fruit is chosen?

9. How many of the 26-letter permutations of the alphabet have no 2 vowels together?

10. How many 10-letter words are there with no 2 adjacent letters the same?

11. How many 10-element subsets of the alphabet have a pair of consecutive letters?

12. How many ways can 5 men and 7 women be seated in a row with no 2 men next to each other?

13. How many ways can 5 men and 7 women be seated at a round table with no 2 men next to each other?

Which of these questions can we answer now?

1. With repetition not allowed and order counting, how many ways are there to select r things from n distinguishable things?

2. With repetition allowed and order counting, how many ways are there to select r things from n distinguishable things?

3. With repetition not allowed and order not counting, how many ways are there to select r things from n distinguishable things?

4. With repetition allowed and order not counting, how many ways are there to select r things from n distinguishable things?

§8. The Birthday Problem

This optional section presents a result that is interesting but nonintuitive.

$1! = 1$	$2! = 2$	$3! = 6$
$4! = 24$	$5! = 120$	$6! = 720$
$7! = 5040$	$8! = 40,320$	$9! = 362,880$

The approximation

$$n! \approx n^n e^{-n} \sqrt{2\pi n} \quad \text{where} \quad e \approx 2.718281828459045\ldots$$

is called *Stirling's Formula*. Stirling's approximation for $n!$ is off by less than 1% for $n > 8$. For example, Stirling's Formula can be used to approximate just how big $\binom{30}{10}$ is—it's about thirty million. An exercise in elementary algebra shows that $4^n/\sqrt{\pi n}$ is a good approximation to $\binom{2n}{n}$.

The result of the following application of Stirling's Formula should be part of everyone's mathematical baggage. Let p_n be the probability that some 2 of n persons picked at random have the same birthday. The questions is, What is the smallest value of n for which $p_n > 1/2$? It is easier to first compute the probability that no 2 of the n persons have the same birthday. Here, we think of a list of the n persons. We then ask, How many ways can each in turn, have a birthday that is different from those above them on the list? We assume that the total number of possible birthdays is 365^n; you may very well make a different assumption.

P(no 2 of n persons picked at random have the same birthday)

$$= \frac{365 \times 364 \times 363 \times 362 \times \cdots \times (366-n)}{365 \times 365 \times 365 \times 365 \times \cdots \times 365}$$

$$= 365! / [(365 - n)!\, 365^n]$$

$$\approx \frac{365^{365} e^{-365} \sqrt{2\pi\, 365}}{(365-n)^{365-n} e^{-(365-n)} \sqrt{2\pi\,(365-n)}\, 365^n}$$

$$= \frac{365^{365-n} e^{-365} \sqrt{2\pi}\, 365^{.5}}{(365-n)^{365-n} e^{-(365-n)} \sqrt{2\pi} (365-n)^{.5}\, 365^n}$$

$$= [365/(365 - n)]^{365.5-n} e^{-n}.$$

Since $(365/342)^{342.5} e^{-23} \approx .493$ and $(365/343)^{343.5} e^{-22} \approx .524$, then we have the surprising result that the probability that at least 2 of 23 persons picked at random have the same birthday is greater than $1/2$. The same formula shows that if there are 41 persons, then there is more than a 90% probability that at least 2 share a birthday.

§9. n Choose r with Repetition

We will use Mississippi problems to find the number of ways to choose r things from n distinguishable things when repetition is allowed. First, be sure that you can answer the following special basic problems.

- How many fence posts in a row are needed if adjacent posts are 2 yards apart and the end posts are 20 yards apart?

- How many dividers are necessary to separate 10 things in a row from one another?

These "bullet problems" appear over and over again in different disguises. There is no reasonable answer to the question, When do I add 1 and when do I subtract 1?

As we consider the following 10 questions, the important thing to notice is that each, after the first, has the same answer as the question above it. Thus we know that the difficult last question has the same answer as the easy first question.

1. How many words are there consisting of 3 D's and 7 U's? (Dividers and Units)

2. How many sequences are there consisting of 3 |'s and 7 ⋆'s? (Bars and Stars)

3. How many sequences are there consisting of 3 +'s and 7 1's?

4. How many nonnegative integer solutions are there to the equation

$$x_1 + x_2 + x_3 + x_4 = 7?$$

The sequence $1\,1\,1 + + 1\,1 + 1\,1$ in #3 corresponds to the solution $3 + 0 + 2 + 2 = 7$ in #4. Likewise, the solution $0 + 0 + 2 + 5 = 7$ in #4 corresponds to the sequence $+ + 1\,1 + 1\,1\,1\,1\,1$ in #3. This correspondence shows that the answer to #4 is the same as the answer to #3.

By thinking of the variables x_1, x_2, x_3, and x_4 as a row of boxes into which we are dropping the 7 1's, we see that #4 and #5 have the same answer. We probably could not have answered #5 without this observation. The important thing to note is that all these questions have the same answer.

5. How many ways are there to put 7 indistinguishable balls into 4 boxes in a row?

6. How many ways are there to put 7 indistinguishable balls into 4 distinguishable boxes?

7. How many ways are there to pass out 7 oranges to 4 children?

8. How many ways are there to pass out 7 indistinguishable cards that say "I CHOOSE YOU" to 4 persons?

9. How many ways are there to choose 7 pieces of fruit from 4 different types available?

10. How many ways are there to choose, with repetition allowed, 7 objects from 4 distinguishable objects?

Now consider the following 10 questions. These are the same questions as above but with "n" and "r" replacing "4" and "7". Again, the important thing to notice is that adjacent questions have the same answer. We get to our goal by starting with a trivial Mississippi problem.

1. How many words are there consisting of $(n-1)$ D's and r U's?

2. How many sequences are there consisting of $(n-1)$ |'s and r ⋆'s?

3. How many sequences are there consisting of $(n-1)$ +'s and r 1's?

4. How many nonnegative integer solutions are there to the equation

$$x_1 + x_2 + x_3 + \cdots + x_{n-1} + x_n = r?$$

5. How many ways are there to put r indistinguishable balls into n boxes in a row?

6. How many ways are there to put r indistinguishable balls into n distinguishable boxes?

7. How many ways are there to pass out r oranges to n children?

8. How many ways are there to put r indistinguishable cards that say "I CHOOSE YOU" into n distinguishable boxes?

9. How many ways are there to choose r pieces of fruit from n different types available?

10. How many ways are there to choose, with repetition allowed, r objects from n distinguishable objects?

The solution to #10 is the same as the solution to #1, which is a very easy Mississippi problem. We notice that the the solution $\frac{(n-1+r)!}{r!(n-1)!}$ can be written $\binom{n+r-1}{r}$ and thus we have the following.

Observation. With repetition allowed, the number of ways to choose r objects from n distinguishable objects is

$$\binom{n+r-1}{r}.$$

The formula in the observation above is well worth remembering, assuming that you know the meaning of $\binom{n+r-1}{r}$. With this observation, we have the contents of Table 1.2.

	Ordered Selections (Permutations)	Unordered Selections (Combinations)
Without Repetition	$\dfrac{n!}{(n-r)!}$	$\binom{n}{r}$
With Repetition (allowed)	n^r	$\binom{n+r-1}{r}$

TABLE 1.2. Selecting r from n Distinguishable Things.

Some comments about notation are in order. We read "$\binom{n+r-1}{r}$" as "n choose r with repetition." Let's be quite clear that in "selecting with repetition" we mean that repetition is *allowed* in making the selection and not that a repetition is *required* in the choice. If we select with repetition 3 pieces of fruit from oranges, bananas, peaches, and apples, we could select 3 oranges or we could select 1 orange, 1 banana, and 1 peach. How many possible possible choices are there in this case? Although each of the answers 20, $\binom{6}{3}$, and $\binom{4+3-1}{3}$ is correct, we prefer, for pedagogical reasons, the last form since this form carries the most information. We also note that, although the language is different, "picking with replacement" is mathematically the same thing as "picking with repetition." Surely, the number of ways to pick 5 cards from a deck, if after each pick the card is replaced in the deck, is $\binom{52+5-1}{5}$. The language for picking ice cream cones seems to require "with repetition" and not "with replacement."

We have noted that the wedge $\wedge$ is an invaluable tool in doing counting problems. We further illustrate the use of wedges in the following problem.

How many ways are there to arrange the letters in

NASHVILLETENNESSEE

with the first N preceding the the first S and with the first E preceding the T?

There are many reasonable ways to attack the problem, depending on what order we put the letters down to form an arrangement. In any case, we begin by making the figure

```
N A S H V I L E T
N   S       L E
N   S         E
              E
              E
```

Let's first consider the problem of arranging only the 3 N's and the 3 S's. Suppose we choose to begin by putting down the 3 S's. We can do this in 1 way. One attack continues with the diagram $_\wedge S_\wedge S_\wedge S_\wedge$, where the 4 wedges indicate possible spaces for the 3 N's under the condition that the left space is chosen at least once. Another attack continues by putting the required N down with the 3 S's to get the diagram $N_\wedge S_\wedge S_\wedge S_\wedge$, with the wedges representing the possible spaces for the 2 remaining N's. (Observe there is no wedge before the N as it would be the same as the wedge after the N, as far as inserting more N's is concerned.) In either case, we see that there are $\binom{4+2-1}{2}$ ways to pick the spaces for the remaining N's and, of course, 1 way to place the N's in these spaces. On the other hand, suppose we choose to begin by putting down the 3 N's first. We can do this in 1 way. This approach continues with the diagram $N_\wedge N_\wedge N_\wedge$, where the 3 wedges indicate possible spaces for the 3 S's. Thus, there are $\binom{3+3-1}{2}$ ways to pick the spaces for the S's and, of course, 1 way to place the S's in these spaces. For a totally different approach, we note that the first letter must be an N and be followed by any arrangement of 2 more N's and 3 S's. This Mississippi problem has the solution $\frac{5!}{2!3!}$. Whether our approach produces $\binom{4+2-1}{2}$, $\binom{3+3-1}{2}$, or $\frac{5!}{2!3!}$, we see that there are 10 ways to put down the 3 N's and 3 S's.

At this point, we have a typical word that, together with wedges, has the diagram $_\wedge N_\wedge N_\wedge S_\wedge S_\wedge S_\wedge N_\wedge$. There are 6 letters and 7 available spaces for the T and E's. We can choose the spaces for the T and 5 E's in $\binom{7+6-1}{6}$ ways, and, then, put in the letters in these spaces in $\binom{5}{1}$ ways, as determined by selecting the position of the T in the diagram $E_\wedge E_\wedge E_\wedge E_\wedge E_\wedge$. We are now in the position of having a typical 12-letter word, together with wedges, as in the diagram $_\wedge N_\wedge N_\wedge S_\wedge T_\wedge S_\wedge E_\wedge E_\wedge S_\wedge N_\wedge E_\wedge E_\wedge E_\wedge$. The 13 wedges here indicate the possible spaces for the remaining 6 letters. We can pick the 6 spaces in $\binom{13+6-1}{6}$ ways, and, then, putting the letters in these spaces is a simple Mississippi problem that can be done in $\frac{6!}{2!}$ ways. The answer to our question is

$$1\binom{3+3-1}{2} 1 \cdot \binom{7+6-1}{6} \binom{5}{1} \cdot \binom{13+6-1}{6} \frac{6!}{2!}.$$

Problems for Class.

1. How many possible outcomes are there if k dice are tossed?

2. How many 4-letter words are there with the letters in alphabetical order?

3. How many ways can 10 men and 7 women sit in a row so that no 2 women are next to each other?

4. How many ways can 10 men and 7 women sit at a round table so that no 2 women are next to each other?

5. How many arrangements of the letters of RECURRENCERELATION have no 2 vowels adjacent?

6. How many arrangements of the letters of RECURRENCERELATION have the vowels in alphabetical order?

7. How many 5-letter words using only A's, B's, C's, and D's are there that do not contain the word BAD?

8. How many ways can 8 persons, including Peter and Paul, sit in a row with Peter and Paul not sitting next to each other?

9. How many ways can 8 persons, including Peter and Paul, sit at a round table with Peter and Paul sitting next to each other?

10. How many ways can 4 persons of each of n nationalities stand in a row with each person standing next to a fellow national?

11. How many ways are there to give each of 5 children 4 of 20 distinguishable toys?

12. How many ways can we partition 18 persons into study groups of 5, 6, and 7?

13. How many ways can we partition 18 persons into 3 study groups of 6?

14. How many arrangements of 7 R's and 11 B's are there such that no 2 R's are adjacent?

15. How many arrangements of the letters in MISSISSIPPI have no 2 I's adjacent?

16. How many nonnegative integer solutions are there to the equation

$$x_1 + x_2 + x_3 + x_4 + x_5 = 67?$$

17. How many ways are there to distribute 30 green balls to 4 persons if Alice and Eve together get no more than 20 and Lucky gets at least 7?

18. How many ways can we pick 18 letters from 7 A's, 8 B's, and 9 C's?

Ten Problems for Homework.

1. How many 4-letter words are there with the letters in alphabetical order?

2. How many 4-letter words are there with no letter repeated and the letters in alphabetical order?

3. How many 5-card poker hands are there with 2 pairs?

4. How many 3-letter words are there with no repeated letter if the middle letter is a vowel?

5. How many arrangements of the letters in MISSISSIPPI have at least 2 adjacent I's?

6. How many possible outcomes are there if a pair of dodecahedral dice, with sides numbered 1 through 12, are tossed?

7. How many ways can we partition 18 persons into study groups of 5, 5, 4, and 4?

8. How many ways can we partition mn distinguishable objects into m piles of n objects each?

9. How many different selections of fruit can be made from 5 oranges and 7 apples?

10. How many different words of at least 1 letter can be made from 3 A's and 3 B's?

You may feel that we should have some special symbol for "n choose r with repetition," as there is for "n choose r." Well, we do. This symbol was not immediately introduced so that we would get into the habit of thinking of "$\binom{n+r-1}{r}$" whenever we said "n choose r with repetition." It would be a good exercise to give the answers to the 5 questions below both with and without the new notation. We define the symbol "$\left\langle \begin{smallmatrix} n \\ r \end{smallmatrix} \right\rangle$," which we read as "$n$ choose r with repetition," by

$$\left\langle {n \atop r} \right\rangle = \binom{n+r-1}{r}.$$

Five Problems for Homework.

1. How many arrangements of the letters in MISSISSIPPI have no P adjacent to an S? (Hint: Although it is the same problem, it is very much easier to consider no S adjacent to a P.)

2. How many nonnegative integer solutions are there to the equation

$$x_1 + x_2 + x_3 + x_4 + x_5 + x_6 = 32?$$

3. How many positive integer solutions are there to the equation

$$x_1 + x_2 + x_3 + x_4 + x_5 + x_6 = 32?$$

4. How many positive integer solutions are there to the inequality

$$x_1 + x_2 + x_3 + x_4 + x_5 + x_6 < 32?$$

(Hint: Consider $x_1 + x_2 + x_3 + x_4 + x_5 + x_6 + x_7 = 32$.)

5. How many nonnegative integer solutions are there to the inequality

$$x_1 + x_2 + x_3 + x_4 + x_5 + x_6 < 32?$$

§10.   Ice Cream Cones—The Double Dip

We know that 5 single dip ice cream cones can be ordered without repetition from 12 available flavors in $\binom{12}{5}$ ways. We also know that 5 single dip ice cream cones can be ordered with repetition allowed from 12 available flavors in $\binom{12+5-1}{5}$ ways. We now face the complications presented by considering the double dip, where 2 flavors may be ordered for 1 cone. We will have to concern ourselves with the true ice cream aficionado who asserts that the order of the scoops on the cone must be considered, as well as those less discerning who consider the order of the scoops to be irrelevant. Instead of counting double dip cones where the order of the scoops is irrelevant, we could alternately count the number of dishes with 2 scoops of ice cream.

1. How many ways can we order with repetition of cones allowed 5 double dip ice cream cones from 12 available flavors when the order of the scoops is taken into consideration?

2. How many ways can we order without repetition of cones 5 double dip ice cream cones from 12 available flavors when the order of the scoops is taken into consideration?

3. How many ways can we order without repetition of cones 5 double dip ice cream cones from 12 available flavors when the order of the scoops is considered irrelevant?

4. How many ways can we order with repetition of cones allowed 5 double dip ice cream cones from 12 available flavors when the order of the scoops is considered irrelevant?

§11.   Block Walking

1. How many ways are there to arrange 10 R's and 6 U's? (Right and Up)

2. How many shortest paths, consisting only of nonoverlapping (except at endpoints) unit segments that are parallel to an axis, are there from the origin in the plane to the point $(10, 6)$? See Figure 1.1, where 2 such paths are shown.

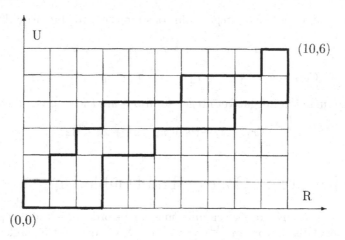

FIGURE 1.1. Block Walking.

3. How many words are there consisting of 3 X's, 6 Y's, and 7 Z's?

4. How many shortest paths, consisting only of nonoverlapping unit segments that are parallel to an axis, are there from the origin in 3-space to the point $(3, 6, 7)$?

5. How many words are there consisting of x X's and y Y's?

6. How many shortest paths, consisting only of nonoverlapping unit segments that are parallel to an axis, are there from the origin in the plane to the point (x, y) with nonnegative integer coordinates? If we mark each such point with the answer, then we should see a definite pattern. If we are doing this on a sheet of paper and if we first rotate our grid so that the origin is at the top before marking the points, then we cannot help but recognize Pascal's Triangle.

7. How many words are there consisting of x X's, y Y's, and z Z's?

8. How many shortest paths, consisting only of nonoverlapping unit segments that are parallel to an axis, are there from the origin in 3-space to the point (x, y, z) with nonnegative integer coordinates?

9. How many rectangles are there in Figure 1.1?

10. How many squares are there in Figure 1.1?

§12. Quickies and Knights

Quickies. It undoubtedly pays off to do this very important exercise more than twice, on different days, until you can zip right through the questions. You should be able to answer these quickies almost in the time it takes to read the questions.

1. How many ways are there to select 5 women from 16 husband–wife couples?

2. How many ways are there to arrange the 7 letters AAABBBB?

3. How many ways are there to arrange, without having 2 C's together, the 12 letters of AAABBBBCCCCC?

4. How many ways are there to seat 10 people in a row?

5. How many ways are there to seat 10 people at a round table?

6. How many distinguishable dominoes are there, (each of the 2 ends of a domino has 0 to 6 dots carved on it)?

7. How many ways can 14 men and 9 women be seated in a row so that no 2 women sit next to each other?

8. How many ways are there to select 10 cans of soda from 4 different brands?

9. How many ways can 22 cans of beer be handed out to 4 people if everyone must get at least 1 can?

10. How many ways are there to pick 9 cans from among 8 cans of each of 57 varieties?

11. How many ways are there to distribute 5 apples and 8 oranges to 6 children?

12. How many ways are there to select some fruit from 5 apples and 8 oranges, taking at least 1 piece?

13. How many nonnegative integers less than a billion have 5 7's?

14. How many 5-letter words can be formed from the alphabet without repeating any letter?

15. How many ways are there to pair off 8 men with 8 women at a dance?

16. How many positive integer solutions are there to the equation $w + x + y + z = 24$?

17. How many ways are there to pick 12 letters from 12 A's and 12 B's?

18. How many ways are there to pick 18 letters from 12 A's and 12 B's?

19. How many ways are there to pick 25 letters from 12 A's, 12 B's and 12 C's?

20. How many ways are there to select a dozen doughnuts chosen from 7 varieties with the restriction that at least 1 doughnut of each variety must be chosen?

21. How many ways are there to assign 50 agents to 5 different countries so that each country get 10 agents?

22. How many ways are there to put 17 red balls into 12 distinguishable boxes with at least 1 ball in each box?

23. How many ways can 9 dice fall?

24. How many ways can 12 pixies (distinguishable, of course) sit at a round table?

25. How many ways are there to arrange 5 C's and 15 R's such that there are at least 2 R's between any 2 C's?

Consider the following problem.

How many ways are there to select 5 integers from $\{1, 2, \ldots, 20\}$ such that the (positive) difference between any 2 of the 5 is at least 3?

This may be a difficult problem until we think of either choosing or rejecting each integer in turn and associate our selection of the 5 integers with a word consisting of 5 C's (for "Choose") and 15 R's (for "Reject") such that there are at least 2 R's between any 2 C's? Aha! This problem is only the last quickie in disguise. To compute the solution, we start with the diagram $_\wedge C_\wedge C_\wedge C_\wedge C_\wedge C_\wedge$, with the 6 wedges indicating places for the insertion of the 15 R's. Each of the inner 4 places must be chosen at least twice, leaving $15 - 8$ arbitrary choices among the 6 places. Thus, we get $\binom{6}{7}$ as our answer. We have changed a fairly hard problem into an easy problem. Once seen, such problems now become quickies.

We will use the same technique in a generalization of a classic problem, which is not a quickie and which you might enjoy trying before reading its solution.

The Knights' Quest. How many ways are there to select 4 knights—just the right number for a quest to rescue a fair maiden—from 15 knights sitting at a round table if no adjacent knights can be chosen?

We assume that Lucky Lancelot Pierre is 1 of the knights at the round table. There are 2 cases: either Lancelot goes on the quest or he does not. First, suppose Lancelot goes on the quest. Now, with Lancelot and hence the 2 knights adjacent to Lancelot eliminated from further consideration in this case, we have 12 knights remaining from which to select 3 with no 2 adjacent. We want to count the number of words consisting of 3 C's (again for "Choose") and 9 R's (again for "Reject") but with no 2 adjacent C's. Putting down the 9 R's first, we see that there are 10 places into which we can insert at most 1 C. There are $\binom{10}{3}$ such words and an equal number of quests having Lancelot as a member. Without Lancelot on the quest, we are looking for the 14-letter words with 4 C's and 10 R's but no adjacent C's. In this case we have $\binom{11}{4}$ possibilities. Hence, our final answer is 450 from

$$\binom{10}{3} + \binom{11}{4}.$$

Let's generalize the problem and suppose that there are k knights and that q are to be chosen for the quest with no adjacent knights chosen. If we calculate the number of quests that have Lancelot as a member and multiply this number by k, then this product equals q times the number that we are seeking. (Not that we are interested, but the product is the number of quests with a designated leader.) Now, with Lancelot and hence the 2 knights adjacent to Lancelot eliminated, we have $k - 3$ knights remaining from which to select $q-1$ with no 2 adjacent. We want to count the number of words consisting of $q - 1$ C's and $(k - 3) - (q - 1)$ R's but with no 2 adjacent C's. There are $\binom{k-q-1}{q-1}$ such words and an equal number of quests having Lancelot as a member. The number we are looking for is $\frac{k}{q}\binom{k-q-1}{q-1}$, or $\frac{k}{k-q}\binom{k-q}{q}$, or

$$\frac{k(k - q - 1)!}{q!(k - 2q)!}.$$

Let's generalize the problem even further. Instead of insisting on only 1 knight between any 2 chosen knights, suppose that there is a minimum number g (for gap) of knights that must be between any 2 chosen knights in the problem. This time, we want to multiply k/q times the number of words with $q - 1$ C's and $(k - (2g + 1)) - (q - 1)$ R's but with any 2 C's having at least g R's between them. In this case, it is much easier to put down the $q - 1$ C's first. The $(q - 1) + 1$ wedges in the diagram $_\wedge C _\wedge C _\wedge C _\wedge \ldots _\wedge C _\wedge C _\wedge$ indicate the places available for inserting R's. We must choose each of the inner $(q - 1) - 1$ places at least g times to meet the condition of having g R's between any 2 C's. This leaves $k - (2g + 1) - (q - 1) - g(q - 2)$ arbitrary choices for as many R's in the q places. Hence, we have a solution $\frac{k}{q}\left\langle\begin{smallmatrix}q\\k-(2g+1)-(q-1)-g(q-2)\end{smallmatrix}\right\rangle$, or $\frac{k}{q}\binom{k-gq-1}{q-1}$, or

$$\frac{k(k - gq - 1)!}{q!(k - (g + 1)q)!}.$$

§13.   The Binomial Theorem

What is the coefficient of $a^i b^j$ in the expansion of the product $(a+b)^n$? We write the product $(a+b)^n$ as the product

$$[a+b][a+b][a+b][a+b]\cdots[a+b][a+b],$$

consisting of n identical binomial factors $a+b$, each within a pair of brackets. In expanding this product, we see that each of the factors within brackets must contribute either an a or a b to each term in the expansion. Collecting like terms, we see that the sum $i+j$ of the exponents of the term $c_{ij} a^i b^j$ in the expansion must be n and that c_{ij} must be just the number of ways of picking the i factors within brackets that contribute an a. So $c_{ij} = \binom{n}{i}$. Likewise, $c_{ij} = \binom{n}{j}$, since this is the number of ways of picking the j factors within brackets that contribute a b. Since $i+j=n$, this merely says that $\binom{n}{i} = \binom{n}{n-i}$.

Observation. For positive integer n,

$$(a+b)^n = \sum_{r=0}^{n} \binom{n}{r} a^r b^{n-r}.$$

This result is called the **Binomial Theorem** and is why the numbers of the form $\binom{n}{r}$ are frequently called **binomial coefficients**.

Set $a = b = 1$ in the Binomial Theorem to get the result

$$\sum_{r=0}^{n} \binom{n}{r} = 2^n.$$

Set $a = -1$ and $b = 1$ in the Binomial Theorem to get the result

$$\binom{n}{0} + \binom{n}{2} + \binom{n}{4} + \cdots = \binom{n}{1} + \binom{n}{3} + \binom{n}{5} + \cdots = 2^{n-1}.$$

There are literally thousands of equations involving the binomial coefficients. We state only 1 more here; as is shown in The Back of the Book, this equality can be proved using a block walking argument:

$$\binom{n}{0}^2 + \binom{n}{1}^2 + \binom{n}{2}^2 + \cdots + \binom{n}{n}^2 = \binom{2n}{n}.$$

Also, by counting the number of ways to select n persons by picking k of m men and $n-k$ of w women, we see that $\sum_{k=0}^{n} \binom{m}{k}\binom{w}{n-k} = \binom{m+w}{n}$. The special case $m = w = n$ gives the desired equation.

§14.   Homework for a Week

1. How many ways are there to put 24 distinguishable flags on 18 distinguishable flagpoles? (Our flagpoles can hold any number of flags. Actually, flagpole problems are only balls-into-boxes problems where order within the boxes is considered.)

2. How many ways are there to put 24 distinguishable flags on 18 distinguishable flagpoles if each flagpole must have at least 1 flag?

3. How many ways can we put r indistinguishable balls into n distinguishable boxes with exactly m of the boxes empty?

4. How many ways are there to distribute 20 oranges to 8 children if the youngest 2 children get the same number of oranges?

5. How many ways are there to distribute 20 distinguishable books to 8 children if the youngest 2 children get the same number of books?

6. How many arrangements of the letters in WISCONSIN have a W adjacent to an I but no 2 consecutive vowels?

7. How many ways are there to distribute 62 indistinguishable white balls and 8 distinguishable numbered balls into 10 distinguishable boxes?

8. How many nonempty collections of balls can be formed from 8 red balls, 9 white balls, and 10 blue balls?

9. How many arrangements of MISSISSIPPI are there in which there is an I adjacent on each side of each P?

10. How many arrangements of MISSISSIPPI are there in which the first I precedes the first S and the first S precedes the first P?

11. In how many arrangements of MISSISSIPPI do both P's precede all the S's?

12. How many ways are there to put 5 indistinguishable red flags, 7 indistinguishable blue flags, and 11 different, distinguishable flags onto 15 distinguishable flagpoles?

13. How many ways are there to seat 8 boys and 13 girls at a round table with no 2 boys adjacent and each girl sitting next to at least 1 boy?

14. How many ways are there to make 95 cents change in 1943 pennies, 1998 pennies, and 2001 quarters?

15. How many election outcomes are there such that exactly 3 of the 5 candidates tie for the most votes from 38 voters?

16. How many ways are there to put n distinguishable balls into n distinguishable boxes so that exactly 2 boxes are empty?

17. How many ways are there to put 8 indistinguishable balls into 6 distinguishable boxes if the first 2 boxes together have at most 4 balls?

18. How many ways are there to put 8 distinguishable balls into 6 distinguishable boxes if the first 2 boxes together have at most 4 balls?

Four Questions for Thought.

Half of the following questions are easy and half are, at this point, nearly impossible. With experience, we will be able to recognize a very hard problem. What is it that makes half of the following very hard?

- How many ways can we put r distinguishable balls into n distinguishable boxes?

- How many ways can we put r indistinguishable balls into n distinguishable boxes?

- How many ways can we put r distinguishable balls into n indistinguishable boxes?

- How many ways can we put r indistinguishable balls into n indistinguishable boxes?

§15.   Three Hour Exams

Practice Exam #1. (Do NOT simplify your answers. You may omit or miss 3 questions without penalty.) **1.** How many 4-letter words are there? **2.** How many 4-letter words have the last letter repeat an earlier letter? **3.** How many ways can 8 individuals sit at a round table? **4.** How many ways can 7 boys and 5 girls line up with all the girls together? **5.** How many ways can 5 1-scoop dishes of ice cream be ordered with repetition if 7 flavors are available? **6.** How many ways can 5 3-scoop dishes of ice cream be ordered with repetition if each scoop can be any 1 of 7 available flavors? **7.** How many ways can 7 oranges and 5 distinguishable toys be distributed to 4 children? **8.** How many ways can 4 A's, 5 B's, and 6 C's be arranged with no B directly following an A? **9.** How many poker hands contain at least 1 card in each suite? **10.** How many ways are there to pick 20 letters from 10 A's, 10 B's, and 10 C's? **11.** How many integer solutions are there to the system $0 < x < y < z < 25$? **12.** How many positive integer solutions are there to the system $x + y + z < 25$? **13.** How many positive integer solutions are there to the system $x+y+z = 25$? **14.** How many ways can we put r indistinguishable balls into s distinguishable boxes? **15.** How many 7 element subsets of the letters of the alphabet have a pair of consecutive letters? **16.** How many arrangements of MISSISSIPPI are there? **17.** How many arrangements of MISSISSIPPI have an S adjacent on each side of the M? **18.** How many arrangements of MISSISSIPPI have both P's precede all the S's? **19.** How many arrangements of MISSISSIPPI have the first S precede the first I and the first I precede the first P? **20.** How many ways can 5 indistinguishable red flags, 7 indistinguishable blue flags, and 11 different, distinct national flags be flown from 14 distinguishable flagpoles?

Practice Exam #2. (Do NOT simplify your answers. You may omit or miss 3 questions without penalty.) **1.** How many ways are there to arrange the letters of MISSISSIPPI? **2.** How many balls must be chosen from 12 red balls, 20 white balls, 7 blue balls, and 8 green balls to be assured that there are 10 balls of the same color? **3.** How many of the 26-letter permutations of the alphabet have no 2 vowels together? **4.** How many 9-element subsets of the letters of the alphabet have no pair of consecutive letters? **5.** How many ways can 6 men and 8 women be seated at a round table with no 2 men next to each other? **6.** How many ways can 9 persons, including Peter and Paul, sit in a row with Peter and Paul not sitting next to each other? **7.** How many 5 letter words are there with no repeated letter if the middle letter is a vowel? **8.** How many arrangements of the letters in MISSISSIPPI have at least 2 adjacent S's? **9.** How many different selections of fruit (including none) can be made from 5 oranges and 7 apples? **10.** How many ways can some (including none and all) of 32 indistinguishable balls be put into 6 distinguishable boxes? **11.** How many ways can we order with repetition of cones allowed 7 double dip ice cream cones from 15 available flavors when the order of the scoops is taken into consideration? **12.** How many ways can we order with repetition of cones allowed 7 double dip ice cream cones from 15 available flavors when the order of the scoops is considered irrelevant? **13.** How many ways are there to distribute 6 apples and 9 oranges to 5 children? **14.** How many ways are there to pick 20 letters from 14 A's and 14 B's? **15.** How many ways are there to put 26 distinguishable flags on 14 distinguishable flagpoles if each flagpole must have at least 1 flag? **16.** How many ways can we put 32 red balls into 12 distinguishable boxes with exactly 3 of the boxes empty? **17.** How many ways are there to distribute 62 indistinguishable white balls and 8 distinguishable numbered balls into 10 distinguishable boxes? **18.** How many ways are there to put 6 indistinguishable red flags, 8 indistinguishable blue flags, and 13 different, distinguishable flags onto 15 distinguishable flagpoles? **19.** How many 10-letter words are there with no 2 adjacent letters the same? **20.** How many arrangements of the word MASSACHUSETTS without any consecutive vowels have an H adjacent to an A?

Practice Exam #3. (Do NOT simplify your answers. You may omit or miss 5 questions without penalty.) **1.** How many ways are there to arrange the letters of MISSISSIPPI? **2.** How many balls must be chosen from 12 red balls, 20 white balls, 7 blue balls, and 8 green balls to be assured that there are 10 balls of the same color? **3.** How many of the 26-letter permutations of the alphabet have no 2 vowels adjacent? **4.** How many ways can 7 distinguishable lions and 4 distinguishable tigers be paraded in line into an arena if no tiger follows directly behind another tiger? **5.** How many ways can 6 men and 8 women be seated at a round table with no 2 men next to each other? **6.** How many ways can 9 persons, including Peter and Paul, sit in a row with Peter and Paul not sitting next to each other? **7.** How many 5-letter words are there in which every 3-letter subword consists of 3 distinguishable letters? **8.** How many arrangements of MISSISSIPPI have the first I precede the first S? **9.** How many different selections of fruit (including none) can be made from 5 oranges and 7 apples? **10.** How many ways can 10 passengers sit in a 10-seat train compartment with 5 seats facing forward and 5 seats facing backward, accommodating 4 who want to face forward and 3 who want to face backward? **11.** How many ways can we order with repetition of cones allowed 7 double dip ice cream cones from 9 available flavors when the order of the scoops is not taken into consideration? **12.** How many ways can 6 speakers be ordered if speaker A must not precede speaker B? **13.** How many ways are there to form 6-element subsets of $\{1, 2, \ldots, 25\}$ such that the largest element is 20? **14.** How many ways are there to form 6-element subsets of $\{1, 2, \ldots, 25\}$ such that the largest element is greater than 20? **15.** How many ways are there to form 6-element subsets of $\{1, 2, \ldots, 25\}$ having no 2 consecutive integers? **16.** How many ways are there to arrange 4 A's, 5 B's, 6 C's, and 7 D's with no consecutive B's? **17.** How many ways are there to arrange 4 A's, 5 B's, 6 C's, and 7 D's with no substring AB occurring? **18.** How many ways are there to seat 7 men and 7 women in a row with men and women alternating? **19.** How many ways are there to seat 7 men and 7 women at a round table with men and women alternating? **20.** How many ways are there to seat 7 inseparable couples at a round table? **21.** How many ways are there to walk a total of 7 blocks north and 4 blocks west while stopping at the shop that is 3 blocks north and 2 blocks west? **22.** How many ways are there to put m distinguishable flags on n indistinguishable flagpoles so that each flagpole has at least 1 flag? **23.** How many ways can 8 indistinguishable red flags, 9 indistinguishable blue flags, and 10 different, distinct national flags be flown from 11 distinguishable flagpoles? **24.** How many ways can 8 indistinguishable red flags, 9 indistinguishable blue flags, and 10 different, distinct national flags be flown from 11 distinguishable flagpoles if each flagpole has at least 1 flag? **25.** How many ways can 8 indistinguishable red flags, 9 indistinguishable blue flags, and 10 different, distinct national flags be flown from 11 distinguishable flagpoles if each flagpole has at least 2 flags?

2

The Principle of Inclusion and Exclusion

§16. Introduction to PIE

Let's begin with some comments about Figure 2.1, which is on the next page. We are looking at properties of subsets A, B, C of some universal set that has t elements. The universal set in the figure is represented by the rectangles. For each set S, we let $|S|$ denote the number of elements in S and say that $|S|$ is the **size** of S. Looking at the top third of the figure, we see a trivial result that is tremendously important. Frequently it is much easier to count the elements in a set by first counting the elements not in the set. For example, counting the 5-letter words that have a vowel is very difficult without first counting the number of 5-letter words that have no vowel. The number of 5-letter words that have a vowel is $26^5 - 21^5$.

Although the middle third of Figure 2.1 may be considered common sense that follows simply from the use of language or is easily proved, the bottom third probably needs some argument. In order to count the number of elements in none of A, B, C, we begin by putting a $+1$ in each of the 8 regions in the figure. Continue by putting a -1 in each of the 4 regions that are within A, then within B, and then within C. Next, we put a $+1$ in each region that lies within at least 2 of the 3 sets. Finally, we put a -1 in the region that lies within all 3 of A, B, C. We should end up with only the region having no elements of A, B, C having a sum of $+1$, with all the others 7 regions having a sum of 0.

We can now understand where the name for PIE comes from. For example, following the procedure given above, we see that an element in $A \cap B \cap C$

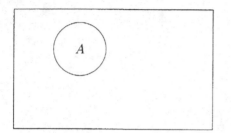

The number of elements
not in A is
$$t - |A|.$$

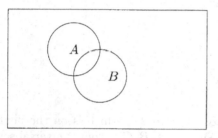

The number of elements
in neither A nor B is
$$t - (|A| + |B|) + |A \cap B|.$$
So, $|A \cup B| = (|A| + |B|) - |A \cap B|$.

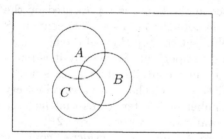

The number of elements
in none of A, B, C is
$$t$$
$$- (|A| + |B| + |C|)$$
$$+ (|A \cap B| + |A \cap C| + |B \cap C|)$$
$$- |A \cap B \cap C|.$$

So, $|A \cup B \cup C|$
$$= (|A| + |B| + |C|) - (|A \cap B| + |A \cap C| + |B \cap C|) + |A \cap B \cap C|.$$

FIGURE 2.1. Venn Diagrams for PIE.

is included once in t, excluded 3 times in $|A| + |B| + |C|$, included 3 times in $|A \cap B| + |A \cap C| + |B \cap C|$, and finally excluded once in $|A \cap B \cap C|$. The total here is 0, as it should be. All this is a bit tedious and it is apparent that using diagrams to go to 5 or more sets is a frightening idea. You may be able to see a pattern emerging and guess the arithmetic result of starting with 4 or more subsets. We will prove this general pattern in the next section. The following problem set is based on only the results stated in Figure 2.1.

PIE Problems I.

1. How many 9-digit (ideal) Social Security numbers are there with repeated digits?

2. How many 4-letter words begin or end with a vowel?

3. How many 4-letter words neither begin nor end with a vowel?

4. How many ways can 25 red balls be put into 3 distinguishable boxes if no box is to contain more than 15 balls?

5. How many ways can 25 red balls be put into 3 distinguishable boxes if no box is to contain more than 10 balls?

6. How many arrangements of the letters in COMBINATORICS have both C's preceding both I's or both O's preceding both I's?

7. How many integer solutions are there to the system

$$x_1 + x_2 + x_3 = 18, \quad 0 \le x_i \le 9?$$

8. How many permutations of the 10 digits contain the sequence 024 or the sequence 456?

9. How many permutations of the 10 digits contain at least 1 of the 3 sequences 123, 345, 567?

10. How many ways are there to roll a die 7 times in a row such that the 5th number shown equals an earlier number?

§17. Proof of PIE

Lemma for PIE. The number of elements in a given set of t elements that are in none of the subsets $A, B, C, \ldots, Z$ of the given set is

$$t$$
$$-\,(|A| + |B| + |C| + \cdots + |Z|)$$
$$+\,(|A \cap B| + |A \cap C| + \cdots + |Y \cap Z|)$$
$$-\,(|A \cap B \cap C| + |A \cap B \cap D| + \cdots + |X \cap Y \cap Z|)$$
$$+\,(|A \cap B \cap C \cap D| + \cdots + |W \cap X \cap Y \cap Z|)$$
$$\vdots$$
$$\pm\,|A \cap B \cap C \cap \cdots \cap Y \cap Z|.$$

Proof. Consider an element of the given set that is in exactly k of the subsets $A, B, C, \ldots, Z$. The number of times this element is counted in the expression in the lemma is, with line by line above corresponding to term by term below, given by

$$1 - \binom{k}{1} + \binom{k}{2} - \binom{k}{3} + \cdots + (-1)^k \binom{k}{k}.$$

By the Binomial Theorem, this quantity is $(1-1)^k$, which is 0 if $k > 0$ and 1 if $k = 0$, which is exactly what is required to prove the lemma. ∎

The theorem PIE, stated below, is an immediate corollary of the lemma. The lemma and the theorem are essentially 2 ways of saying the same thing. In fact, some refer to our lemma as the Principle of Inclusion and Exclusion and call our theorem a corollary.

PIE (*The Theorem*). The number of elements in 1 or more of the finite sets $A, B, C, \ldots, Z$ is

$$+\,(|A| + |B| + |C| + \cdots + |Z|)$$
$$-\,(|A \cap B| + |A \cap C| + \cdots + |Y \cap Z|)$$
$$+\,(|A \cap B \cap C| + |A \cap B \cap D| + \cdots + |X \cap Y \cap Z|)$$
$$-\,(|A \cap B \cap C \cap D| + \cdots + |W \cap X \cap Y \cap Z|)$$
$$\vdots$$
$$\pm\,|A \cap B \cap C \cap \cdots \cap Y \cap Z|.$$

We restate PIE in a different notation that is sometimes useful. For reference, a couple generalizations of PIE are stated without proof below.

Let S_k denote the sum of the sizes of all k-tuple intersections of given sets A_1, A_2, A_3, ... , A_n. Note that in many problems S_k is the sum of $\binom{n}{k}$ identical terms. We have

$$S_1 = |A_1| + |A_2| + |A_3| + \cdots + |A_n|,$$
$$S_2 = |A_1 \cap A_2| + |A_1 \cap A_3| + \cdots + |A_{n-1} \cap A_n|,$$
$$S_3 = |A_1 \cap A_2 \cap A_3| + |A_1 \cap A_2 \cap A_4| + \cdots + |A_{n-2} \cap A_{n-1} \cap A_n|,$$
$$S_4 = |A_1 \cap A_2 \cap A_3 \cap A_4| + \cdots + |A_{n-3} \cap A_{n-2} \cap A_{n-1} \cap A_n|$$

$$\vdots$$

$$S_n = |A_1 \cap A_2 \cap A_3 \cap \cdots \cap A_{n-1} \cap A_n|.$$

PIE and Two Generalizations. Let S_k denote the sum of the sizes of all k-tuple intersections of given sets A_1, A_2, A_3, ... , A_n. The number of elements in at least 1 of the given sets is

$$S_1 - S_2 + S_3 - S_4 + \cdots + (-1)^{k-1} S_k + \cdots + (-1)^{n-1} S_n.$$

The number of elements in exactly r of the n given sets is

$$S_r - \binom{r+1}{r} S_{r+1} + \binom{r+2}{r} S_{r+2} - \cdots + (-1)^k \binom{r+k}{r} S_{r+k} + \cdots + (-1)^{n-r} \binom{n}{r} S_n.$$

The number of elements in at least r of the n given sets is

$$S_r - \binom{r}{r-1} S_{r+1} + \binom{r+1}{r-1} S_{r+2} - \cdots + (-1)^k \binom{r+k-1}{r-1} S_{r+k} + \cdots + (-1)^{n-r} \binom{n-1}{r-1} S_n.$$

PIE Problems II.

1. How many 10-letter words do not contain all the vowels?

2. How many ways can 40 red balls be put into 4 distinguishable boxes with at most 15 balls in each box?

3. How many ways can 11 distinguishable balls be put into 6 distinguishable boxes with at least 1 box empty?

4. How many integer solutions are there to the system

$$x_1 + x_2 + x_3 + x_4 = 100, \quad 0 \le x_i \le 30?$$

5. How many ways can 10 coins be selected from a bag of 7 pennies, 8 nickels, and 7 quarters?

6. How many ways can r distinguishable balls be placed in n distinguishable boxes with no box empty?

§18.   Derangements

The Hatcheck Problem. How many ways can a hatcheck girl hand back the n hats of n gentlemen, 1 to each gentleman, with no man getting his own hat?

We can tell from the language of the Hatcheck Problem that it is a very old problem. It may be old, but it is great. In any case, an arrangement of the first n positive integers in which no integer is in its "natural" position is called a **derangement** of the n integers. The number of derangements of $\{1, 2, 3, \ldots, n\}$ is denoted by D_n. In general, the **derangement** of a string of distinct elements is an arrangement of the the elements such that no element appears in its original position. The solution to the Hatcheck Problem is D_n. We wish to find a formula for D_n.

We let A_i be the set of arrangements of $\{1, 2, 3, \ldots, n\}$ for which i is in the ith position. We need to count the arrangements that are in none of the sets A_i. By the Lemma for PIE, we then calculate that

$$D_n$$

$$= n! - \binom{n}{1}(n-1)! + \binom{n}{2}(n-2)! - \binom{n}{3}(n-3)! + \cdots + (-1)^n \binom{n}{n} 0!$$

$$= n! \left[1 - \frac{1}{1!} + \frac{1}{2!} - \frac{1}{3!} + \frac{1}{4!} - \frac{1}{5!} + \cdots + (-1)^n \frac{1}{n!} \right].$$

Recalling (or taking it as a definition) that $e^x = \sum_{k=0}^{\infty} \frac{1}{k!} x^k$ for real number x, then

$$\frac{1}{e} = e^{-1} = \sum_{k=0}^{\infty} (-1)^k \frac{1}{k!} = 1 - \frac{1}{1!} + \frac{1}{2!} - \frac{1}{3!} + \frac{1}{4!} - \frac{1}{5!} + \cdots + (-1)^k \frac{1}{k!} + \cdots$$

We see that $D_n/n!$ is very close to the number $1/e$. In fact, D_n is the integer closest to $n!/e$ for all n. So $D_n \approx n! \times 0.367879441171442 \ldots$.

If letters, 1 addressed to each of some people, are passed out at random, 1 to each of these people, what is the probability that no person gets the letter addresses to them? This is, of course, the Hatcheck Problem, except that the number of persons is not given. What is absolutely astounding is that the probability is approximately $1/e$—no matter how many people there are!

We compare $D_n/n!$ and $D_{n-1}/(n-1)!$ to see how little the difference is. From

$$\frac{D_n}{n!} = 1 - \frac{1}{1!} + \frac{1}{2!} - \frac{1}{3!} + \frac{1}{4!} - \frac{1}{5!} + \cdots + (-1)^{n-1} \frac{1}{(n-1)!} + (-1)^n \frac{1}{n!}$$

we subtract

$$\frac{D_{n-1}}{(n-1)!} = 1 - \frac{1}{1!} + \frac{1}{2!} - \frac{1}{3!} + \frac{1}{4!} - \frac{1}{5!} + \cdots + (-1)^{n-1} \frac{1}{(n-1)!}$$

to get

$$\frac{D_n}{n!} - \frac{D_{n-1}}{(n-1)!} = (-1)^n \frac{1}{n!}.$$

This is all very interesting. However, somewhat serendipitiously, we have stumbled on something even more interesting. We multiply both sides of our last equation by $n!$ to get a wonderful recurrence relation:

$$D_n = nD_{n-1} + (-1)^n \text{ for } n \geq 2.$$

The following bit of trivia may be of interest. Occasionally one sees the symbol "n_i", which is read "n subfactorial" and is defined by

$$n_i = \sum_{k=0}^{n} (-1)^k \binom{n}{k} (n-k)! = n! \sum_{k=0}^{n} \frac{(-1)^k}{k!}.$$

In other words, we have $n_i = D_n$.

PIE Problems III.

1. How many ways can 20 balls be chosen from 12 black balls, 12 white balls, 8 orange balls, and 8 green balls?

2. How many 12-term sequences of digits do not contain all of the 10 digits?

3. How many ways can 3 couples sit at a round table with no 2 from any couple sitting next to each other?

4. How many arrangements of the letters in COMBINATORICS have both C's preceding both O's, both O's preceding both I's, or both I's preceding both C's?

5. How many arrangements of the letters in COMBINATORICS have both C's preceding both O's, both O's preceding both I's, or both I's preceding the S?

PIE Problems IV.

1. How many ways can the 8 integers 1,2,3, ... ,8 be rearranged with i never immediately followed by $i+1$?

2. How many ways are there for 8 children on a merry-go-round to change places so that somebody new is in front of each child? (The horses are indistinguishable and in a circle.)

3. How many ways are there to distribute 37 distinguishable books to 23 boys if each boy must get at least 1 book?

4. How many integer solutions are there to the system

$$x_1 + x_2 + x_3 + x_4 = 40,$$
$$1 \le x_1 \le 5, \quad 2 \le x_2 \le 7, \quad 3 \le x_3 \le 9, \quad 5 \le x_4 ?$$

5. How many ways can 10 distinguishable balls be put into 3 indistinguishable boxes?

6. What percentage of the randomly arranged 26 letters of the alphabet have at least 1 letter in its natural position?

7. Prove the recurrence relation

$$D_n = (n-1)[D_{n-1} + D_{n-2}] \quad \text{if } n > 2.$$

§19. Partitions

A **partition** of a positive integer r is a collection of positive integers whose sum is r. If there are n terms in the sum, then r is said to be **partitioned into n parts**. For example, 5 has 7 partitions:

$$5; \ 4+1, \ 3+2; \ 3+1+1, \ 2+2+1; \ 2+1+1+1; \ 1+1+1+1+1.$$

Note that $3+2$ and $2+3$ are not considered different partitions of 5. Also, 5 can be partitioned into 1 part in 1 way; 5 can be partitioned into 2 parts in 2 ways; 5 can be partitioned into 3 parts in 2 ways; 5 can be partitioned into 4 parts in 1 way; and 5 can be partitioned into 5 parts in 1 way. We let $\Pi(r, n)$ denote the number of partitions of r into n parts. So $\Pi(r, n)$ is the number of ways of putting r indistinguishable balls (or 1's) into n indistinguishable boxes with no box empty. We will be using this balls-and-boxes idea below. We let $\Pi(r)$ denote the total number of partitions of r. Hence, $\Pi(r)$ is the number of ways of of putting r indistinguishable balls into r indistinguishable boxes. So,

$$\Pi(r) = \sum_{j=1}^{r} \Pi(r, j) = \Pi(r + r, r) = \Pi(2r, r),$$

where the middle equal sign comes from thinking of running to the nearest hardware store and borrowing r additional balls, each indistinguishable from the r that we have already put into the boxes. We then put 1 of the borrowed r balls into each of the r boxes. Now we are sure that no box is empty. We thus have a partition of $2r$ into r parts. Conversely, if we have $2r$ red balls in r indistinguishable boxes, with no box empty, then we can take 1 ball from each box to obtain a partition of r and to return the r borrowed balls to the hardware store.

$r\backslash n$	1	2	3	4	5	6	7	8	9	10	11	12
1	1											
2	1	1										
3	1	1	1									
4	1	2	1	1								
5	1	2	2	1	1							
6	1	3	3	2	1	1						
7	1	3	4	3	2	1	1					
8	1	4	5	5	3	2	1	1				
9	1	4	7	6	5	3	2	1	1			
10	1	5	8	9	7	5	3	2	1	1		
11	1	5	10	11	10	7	5	3	2	1	1	
12	1	6	12	15	13	11	7	5	3	2	1	1
13	1	6	14	18	18	14	11	7	5	3	2	1
14	1	7	16	23	23	20	15	11	7	5	3	2
15	1	7	19	27	30	26	21	15	11	7	5	3
16	1	8	21	34	37	35	28	22	15	11	7	5
17	1	8	24	39	47	44	38	29	22	15	11	7
18	1	9	27	47	57	58	49	40	30	22	15	11
19	1	9	30	54	70	71	65	52	41	30	22	15
20	1	10	33	64	84	90	82	70	54	42	30	22

TABLE 2.1. $\Pi(r,n)$: Partitions of r into n Parts.

We observe that $\Pi(r,n)$ is defined by

$$\Pi(r,n) = 0 \text{ if } n > r, \quad \Pi(r,r) = \Pi(r,1) = 1, \quad \text{and}$$

$$\Pi(r,n) = \sum_{k=1}^{n} \Pi(r-n,k) = \Pi(r-1,n-1) + \Pi(r-n,n),$$

where the last equal sign follows by considering the 2 cases of whether there is a box with exactly 1 ball or there is no box with exactly 1 ball.

Although neither of the 2 recurrence relations given above is anywhere as nice as that for building Pascal's Triangle, we can still use either to compute the entries of the table for $\Pi(r,n)$ that are shown in Table 2.1. For example, from the formula $\Pi(r,n) = \Pi(r-1,n-1) + \Pi(r-n,n)$ we calculate $\Pi(20,6) = 90$ from $\Pi(19,5) + \Pi(14,6) = 70 + 20$; in the table the 90 in the bottom row is the sum of the 70 immediately above to the left and the 20 above the 90 in the same column.

If you want a nice closed formula for $\Pi(r,n)$ you will have to do something that has not been done before.

§20. Balls into Boxes

We are now prepared to answer the basic questions concerning putting balls into boxes. We shall consider 8 cases, depending on whether the balls are distinguishable or indistinguishable, whether the boxes are distinguishable or indistinguishable, and whether some of the boxes can be empty or not. The order in which we consider these is as follows:

1. How many ways can r indistinguishable balls be put into n indistinguishable boxes with no box empty?

This is easy, in a sense, but only because we have just been considering this sort of thing in the previous section. The best answer we can give is $\Pi(r, n)$.

2. How many ways can r indistinguishable balls be put into n indistinguishable boxes?

By borrowing n balls (again), we see that the solution is $\Pi(r + n, n)$.

3. How many ways can r indistinguishable balls be put into n distinguishable boxes?

This we know well: $\left\langle {n \atop r} \right\rangle$.

4. How many ways can r indistinguishable balls be put into n distinguishable boxes with no box empty?

This is just $\left\langle {n \atop r-n} \right\rangle$, which equals $\binom{r-1}{n-1}$.

5. How many ways can r distinguishable balls be put into n distinguishable boxes?

This is the one we could do in high school: n^r.

6. How many ways can r distinguishable balls be put into n distinguishable boxes with no box empty?

Let A_i denote those placements of r distinguishable balls into n distinguishable boxes such that the ith box is empty. Since we want to count the number of placements that are in none of the A_i, we use the Lemma for PIE to get

$$\sum_{k=0}^{n}(-1)^k \binom{n}{k}(n-k)^r.$$

(Those who have studied series in calculus will be able to expand $(e^x - 1)^n$ into a power series in x, after first using the Binominal Theorem, to see that our solution is the coefficient of $x^r/r!$ in the series expansion of $(e^x - 1)^n$.)

7. How many ways can r distinguishable balls be put into n indistinguishable boxes with no box empty?

Once the balls are in the boxes, the boxes become distinguishable by their contents. (Hey! That's a neat trick!) Hence, we need only divide the previous answer by $n!$ to get the solution

$$\frac{1}{n!}\sum_{k=0}^{n}(-1)^k \binom{n}{k}(n-k)^r.$$

This number is denoted by $\left\{{r\atop n}\right\}$ and is called a **Stirling number of the second kind**. Note that $\left\{{r\atop 1}\right\} = \left\{{r\atop r}\right\} = 1$ and for $1 < n < r$ we have

$$\left\{{r\atop n}\right\} = \left\{{r-1\atop n-1}\right\} + n\left\{{r-1\atop n}\right\}.$$

This last equality comes from considering the 2 cases of whether the rth ball is in a box by itself or not. This recurrence relation can be used to form a table of Stirling numbers of the second kind or to compute a particular value with a computer.

We digress only because inquiring minds want to know about the implied Stirling numbers of the first kind. Stirling number $\left\{{r\atop n}\right\}$ counts the number of ways to partition a set of r things into n nonempty subsets. **Stirling number of the first kind** $\left[{r\atop n}\right]$ counts the number of of ways to partition a set of r things into n nonempty subsets and then arrange each subset around a circle. That is, $\left[{r\atop n}\right]$ answers the question, How many ways are there to seat r persons at n indistinguishable round tables with at least 1 person at each table? We can read "$\left\{{r\atop n}\right\}$" as "$r$ subset n" and we can read "$\left[{r\atop n}\right]$" as "r cycle n." Evidently $\left[{r\atop 1}\right] = (r-1)!$, $\left[{r\atop r}\right] = 1$, and for $1 < n < r$ we have

$$\left[{r\atop n}\right] = \left[{r-1\atop n-1}\right] + (r-1)\left[{r-1\atop n}\right].$$

This last equation comes from considering the 2 cases of whether Lucky Pierre sits at a table alone or not. (If not, he can be squeezed to the right of any of the other $r-1$ persons.) Now you know.

8. How many ways can r distinguishable balls be put into n indistinguishable boxes?

The best we can do here is consider all the cases of how many boxes have at least 1 ball in them. So we get

$$\sum_{k=1}^{n}\left\{{r\atop k}\right\}.$$

Our results are summarized in 2 tables. Results #6, 7, 4, and 1 form Table 2.2, while results #5, 8, 3, and 2 form Table 2.3.

Optional Problem.

How many ways can 6 persons sit at 3 indistinguishable round tables if

1. there is at least 1 person at each table?

2. empty tables are allowed?

Balls \ Boxes	Distinguishable	Indistinguishable
Distinguishable	$\sum_{k=0}^{n}(-1)^k\binom{n}{k}(n-k)^r$ $= n!\left\{{r \atop n}\right\}$	$\left\{{r \atop n}\right\}$ $= \frac{1}{n!}\sum_{k=0}^{n}(-1)^k\binom{n}{k}(n-k)^r$
Indistinguishable	$\binom{r-1}{n-1}$	$\Pi(r,n)$ where $\Pi(r,n)=0$ if $n>r$, $\Pi(r,r)=\Pi(r,1)=1$, and $\Pi(r,n)=\sum_{k=1}^{n}\Pi(r-n,k)$

TABLE 2.2. Put r Balls into n Boxes with No Box Empty.

Balls \ Boxes	Distinguishable	Indistinguishable
Distinguishable	n^r	$\sum_{k=1}^{n}\left\{{r \atop k}\right\}$
Indistinguishable	$\binom{n+r-1}{r}$	$\Pi(r+n,n)$

TABLE 2.3. Put r Balls into n Boxes.

§21. A Plethora of Problems

1. How many ways are there to map a set M of m elements into a set N of n elements such that each element of N is the image of at least 1 element of M? (Such mappings are said to be **onto**.)

2. How many ways can a coin be flipped 25 times to get only 8 tails but no run of 5 or more heads?

3. How many ways are there to arrange the 12 integers 1, 2, 3, ... , 12 such that none of 1, 2, 3, 4, 5, and 6 is in its natural position?

4. How many ways are there to seat the persons in 12 couples at a round table such that no 2 of any couple sit next to each other?

5. How many ways can 5 pairs of persons be repaired so that no pair is maintained?

6. How many ways can each of 5 persons be given a right glove and a left glove from 6 distinguishable pairs of gloves but with no person getting a pair of gloves?

7. How many ways can each of 5 persons be given 2 gloves from 6 distinguishable pairs of gloves—that's 12 distinguishable gloves —but with no person getting a pair of gloves?

8. How many ways are there to rearrange the 8 camels behind the lead camel in a caravan so that each of the 8 camels has a different camel in front of it?

9. How many arrangements of RECURRENCERELATION have the first R preceding the first C or the first E preceding the first N?

10. How many arrangements of RECURRENCERELATION have either the first R preceding the first C or the first E preceding the first N?

11. How many arrangements of RECURRENCERELATION have neither the first R preceding the first C nor the first E preceding the first N?

12. How many arrangements of RECURRENCERELATION have the first R preceding the first C and the first E preceding the first C?

13. How many arrangements of RECURRENCERELATION have the first R preceding the first C and the first R preceding the first E?

14. How many arrangements of RECURRENCERELATION have all R's preceding all E's, all E's preceding all C's, and all C's preceding all N's?

15. How many arrangements of RECURRENCERELATION have all R's preceding all E's, all E's preceding all C's, or all C's preceding all N's?

16. How many arrangements of RECURRENCERELATION have neither all R's preceding all E's, all E's preceding all C's, nor all C's preceding all N's?

17. How many arrangements of RECURRENCERELATION have the first R preceding the first E, the first E preceding the first C, and the first N preceding the first C?

18. How many arrangements of RECURRENCERELATION have the first R preceding the first E, the first E preceding the first C, or the first N preceding the first C?

§22. Eating Out

The **Ménage Problem** (Problème des Ménages) asks, How many ways can n husband–wife couples be seated at a round dinner table with men and women alternating but no one sits next to their spouse?

You may wish to try this nontrivial problem for yourself before reading any further.

We consider only seatings where husbands and wives alternate. Let A_i be seatings with couple i sitting together. We want to count the seatings that are in none of the A_i. So, we want

$$t - S_1 + S_2 - S_3 + S_4 + \cdots + (-1)^k S_k + \cdots + (-1)^n S_n$$

from the Lemma for PIE, with $t = (n-1)!\, n!$. The $\binom{n}{k}$ terms in S_k are all equal. We suppose Adam and Eve are the first couple. We compute

$$|A_1 \cap A_2 \cap A_3 \cap \cdots \cap A_k|$$

by placing persons around the table but, as usual, nobody sitting down until all are at the table as follows:

(1) We place Adam and Eve at the table.   2 ways
There are $k-1$ additional couples, $n-k$ additional
males, and $n-k$ additional females to place.
(2) We place the $n-k$ additional males and the $(n-k)!^2$ ways
$n-k$ additional females, realizing that
sexes must alternate here if they are to
alternate after the couples are inserted.
(3) Since Adam and Eve may not be split, there $\left\langle \begin{matrix} 2n-2k+1 \\ k-1 \end{matrix} \right\rangle$ ways
are $2(n-k)+1$ spaces for the $k-1$ couples
to fit into. We choose the spaces.
(4) We place the $k-1$ couples in the spaces, $(k-1)!$ ways
remembering that the sexes alternate.
So $S_k =$

$$\binom{n}{k}\left[2(n-k)!^2 \left\langle \begin{matrix} 2n-2k+1 \\ k-1 \end{matrix} \right\rangle (k-1)! \right] = 2\binom{n}{k}(n-k)!^2 \frac{(2n-k-1)!}{(2n-2k)!}.$$

Evaluating the right-hand side when $k = 0$, we get $(n-1)!\,n!$, which is t. Hence, our solution to the Ménage Problem is

$$2\sum_{k=0}^{n}(-1)^k \binom{n}{k}(n-k)!^2 \frac{(2n-k-1)!}{(2n-2k)!},$$

or

$$2n! \sum_{k=0}^{n}(-1)^k \frac{(n-k)!}{2n-k}\binom{2n-k}{k}.$$

If the chairs are not all indistinguishable from one another, as is sometimes the case in the statement of the Ménage Problem, then we must multiply the solution above by $2n$ to account for rotations of our seatings.

Related Problem.

How many ways are there to permute $\{1, 2, \ldots, n\}$ such that none of the following $2n$ conditions is satisfied?

$$1 \text{ is } 1^{\text{st}}; \quad 1 \text{ is } 2^{\text{nd}};$$
$$2 \text{ is } 2^{\text{nd}}; \quad 2 \text{ is } 3^{\text{rd}};$$
$$3 \text{ is } 3^{\text{rd}}; \quad 3 \text{ is } 4^{\text{th}};$$
$$\cdots$$
$$n-1 \text{ is } (n-1)^{\text{st}}; \quad n-1 \text{ is } n^{\text{th}};$$
$$n \text{ is } n^{\text{th}}; \quad n \text{ is } 1^{\text{st}}.$$

Eating Out Problems.

1. Suppose n persons have both lunch and dinner together. For a given seating arrangement at lunch, verify the displayed solution to the question, How many ways are there to seat these persons for dinner such that no one has the same neighbor on the right at both meals, under the specified condition?

   (a) Both meals are at a counter.

   $$D_n + D_{n-1}$$

   (b) Both meals are at a round table.

   $$D_{n-1} - D_{n-2} + D_{n-3} - D_{n-4} + \cdots \pm D_2$$

   (c) Lunch is at a round table and dinner is at a counter.

   $$nD_{n-1}$$

   (d) Lunch is at a counter and dinner is at a round table.

   $$D_{n-1}$$

2. How many ways are there to seat n wife–husband couples in the manner specified?

   (a) At a round table with no husband and wife diametrically opposite each other.

   (b) At a round table with no wife next to her husband on his right.

   (c) At a round table with no wife next to her husband.

   (d) At a round table with no wife sitting next to her husband on his right if men and women alternate.

   (e) At a lunch counter with no wife next to her husband on his right.

   (f) At a lunch counter with no wife next to her husband.

   (g) At a lunch counter with no wife next to her husband on his right if men and women alternate.

(h) At a lunch counter with no wife next to her husband if men and women alternate.

(i) At a lunch counter with no wife anywhere to the right of her husband.

(j) At a lunch counter with no wife anywhere to the right of her husband if men and women alternate.

3. The following 2 problems are presented for reading only. Although they differ from problems in #1 above only by the deletion of "on the right," they are of a totally different order of difficulty.

(a) Suppose n persons have both lunch and dinner together at a counter. For a given seating arrangement at lunch, how many ways are there to seat these persons for dinner such that no one has the same neighbor at both meals?

(b) Suppose n persons have both lunch and dinner together at a round table. For a given seating arrangement at lunch, how many ways are there to seat these persons for dinner such that no one has the same neighbor at both meals?

3
Generating Functions

§23.   What Is x?

We are familiar with *polynomial equations* such as $x^2 - 4x + 3 = 0$. Here, x is a symbol denoting some real number, called an *unknown*, and our task is usually to try to find its value.

We are familiar with equations for *polynomial functions* such as f where $f(x) = 3 - 4x + x^2$. The equation tells us how f acts on x where x is a *variable*, a symbol denoting an arbitrary real number. So $3 - 4x + x^2$ is the image of x under f for each real number x.

We are probably, perhaps unknowingly, also familiar with *polynomials* in x such as $3 - 4x + x^2$. Here, x is an element not in the set $\mathbb{R}$ of real numbers and is called an **indeterminate**. In this context, x is simply x; and "$x = 2$" is as absurd as "$3 = 2$." The set of all real polynomials in x forms what algebraists call a commutative ring and is denoted by $\mathbb{R}[x]$. This is a subset of $\mathbb{R}[[x]]$, which consists of all real **formal power series** in the indeterminate x. (We read "$\mathbb{R}[[x]]$" as "R double square brackets x.") The elements of $\mathbb{R}[[x]]$ look like

$$\sum_{r=0}^{\infty} a_r x^r \text{ with } a_r \in \mathbb{R}, \text{ where}$$

$$\sum_{r=0}^{\infty} a_r x^r = a_0 + a_1 x + a_2 x^2 + a_3 x^3 + a_4 x^4 + \cdots.$$

Note that we have already implicitly declared that $x^0 = 1$ and $x^1 = x$. Addition in $\mathbb{R}[[x]]$ is defined, as expected, by

$$\left[\sum_{i=0}^{\infty} a_i x^i\right] + \left[\sum_{j=0}^{\infty} b_j x^j\right] = \left[\sum_{k=0}^{\infty} c_k x^k\right] \text{ where } c_k = a_k + b_k.$$

Multiplication is also defined as expected by what is usually called the *Cauchy product*

$$\left[\sum_{i=0}^{\infty} a_i x^i\right] \left[\sum_{j=0}^{\infty} b_j x^j\right] = \left[\sum_{k=0}^{\infty} c_k x^k\right] \text{ where } c_k = \sum_{i+j=k} a_i b_j.$$

This is, after all, how we have always multiplied polynomials and series before we thought deeply about them. For our purposes, the fantastically wonderful thing about formal power series as opposed to the power series we may have seen in calculus is that we never have to worry about whether a formal series converges or not, since the question of convergence does not make any sense because x is an indeterminate and not a variable. For example, the existence of the element $\sum_{r=0}^{\infty} r^r x^r$ in $\mathbb{R}[[x]]$ does not disturb us at all—we don't care how big those coefficients get. To ask whether this element converges or not makes as much sense as asking whether the element 10 converges. Although $\mathbb{R}[x]$ and $\mathbb{R}[[x]]$ have all kinds of interesting properties that are studied in advanced algebra classes, about all we are going to need to know is how to add and multiply their elements.

This introduction to $\mathbb{R}[[x]]$ has been intuitive. There is some polishing that needs to be done in order to be rigorous. We are repeating history here. If you are happy with the introduction, then you can skip the next section. If you see some of the problems or have had your curiosity piqued, then continue on to the next section. There, an outline of a rigorous development of $\mathbb{R}[[x]]$ is given in a journey that will comfortably return us to this point.

§24. An Algebraic Excursion to $\mathbb{R}[[x]]$

A critical eye might notice an immediate problem with our intuitive introduction of $\mathbb{R}[[x]]$ in the previous section. In the first place, the notation $\sum_{r=0}^{\infty} a_r x^r$ for the elements in $\mathbb{R}[[x]]$ already involves addition and multiplication of elements in $\mathbb{R}[[x]]$, even before the addition and multiplication are defined by the expected formulas. Even so, the notation forces us to make correct conclusions, whose only basis is intuition. The rest of this section is devoted to an outline of a rigorous development of $\mathbb{R}[[x]]$.

We begin at the beginning, all over again. The elements of $\mathbb{R}[[x]]$ are defined to be the infinite sequences $(a_0, a_1, a_2, a_3, \ldots)$ of elements from

$\mathbb{R}$. So

$$(a_0, a_1, a_2, a_3, \dots) = (b_0, b_1, b_2, b_3, \dots) \text{ iff } a_i = b_i \text{ for all } i.$$

(Not to worry, x will appear latter.) We define addition $\oplus$ coordinatewise on $\mathbb{R}[[x]]$ by

$$(a_0, a_1, a_2, a_3, \dots) \oplus (b_0, b_1, b_2, b_3, \dots)$$
$$= (c_0, c_1, c_2, c_3, \dots) \text{ where } c_i = a_i + b_i \text{ for all } i,$$

distinguishing the addition in $\mathbb{R}[[x]]$ by $\oplus$ from the garden variety of addition on real numbers denoted by $+$. We define multiplication $\otimes$ on $\mathbb{R}[[x]]$ by

$$(a_0, a_1, a_2, a_3, \dots) \otimes (b_0, b_1, b_2, b_3, \dots)$$
$$= (c_0, c_1, c_2, c_3, \dots) \text{ where } c_k = \sum_{i+j=k} a_i b_j \text{ for all } k,$$

with the Cauchy product in mind. The summation sign refers to addition in $\mathbb{R}$ and causes no problem. Because addition and multiplication are commutative in $\mathbb{R}$, it follows that addition and multiplication are commutative in $\mathbb{R}[[x]]$. That is, if f and g are elements of $\mathbb{R}[[x]]$, then $g \oplus f = f \oplus g$ and $g \otimes f = f \otimes g$. Next, an algebraists would prove all the other properties of a commutative ring: for example, that addition and multiplication are associative and that the distributive laws hold. This is tedious but straightforward. Next we *define* x to be $(0, 1, 0, 0, 0, \dots)$. Why on earth would we do such a thing as introduce x? Well, the only honest answer is that, frankly, we like x. We want to get back to the familiar notation at the beginning of the first section of this chapter. So, $x^2 = (0, 0, 1, 0, 0, 0, \dots)$, $x^3 = (0, 0, 0, 1, 0, 0, 0, \dots)$, etc. Then we embed $\mathbb{R}$ in $\mathbb{R}[[x]]$ with $a \to (a, 0, 0, 0, \dots)$, and we write a in place of $(a, 0, 0, 0, \dots)$. (This is like embedding $\mathbb{R}$ into $\mathbb{C}$, the set of complex numbers, with $a \to a + 0i$.) This is reasonable since the power series 0 and 1 are the zero and unity of $\mathbb{R}[[x]]$, meaning that $g \oplus 0 = 0 \oplus g = g$ and $g \otimes 1 = 1 \otimes g = g$ for all g in $\mathbb{R}[[x]]$. Further, it is customary to write 1 for x^0 and to write x^k for $x \otimes x \otimes x \cdots \otimes x \otimes x$ when there are k factors in the product. Since

$$(a_0, a_1, a_2, a_3, \dots) = (a_0 \otimes x^0) \oplus (a_1 \otimes x^1) \oplus (a_2 \otimes x^2) \oplus (a_3 \otimes x^3) \oplus \cdots$$

and since

$$(a_i b_j, 0, 0, 0, \dots) \otimes x^{i+j} = (a_i \otimes x^i) \otimes (a_j \otimes x^j),$$

where the a_i and b_j on the left side of the equations are real numbers and on the right are power series, we feel free to change the notation for addition and multiplication and write $a_0 + a_1 x + a_2 x^2 + a_3 x^3 + \cdots$ in place of $(a_0, a_1, a_2, a_3, \ldots)$. If $a_i = 0$ for all but a finite number of i, then we call the series a **polynomial** and denote the set of polynomials by $\mathbb{R}[x]$. Finally, we usually omit writing down terms of the form $0x^i$ with $i > 0$. We are back to the familiar notation for $\mathbb{R}[x]$ and $\mathbb{R}[[x]]$, but with everything on a rigorous basis.

§25. Introducing Generating Functions

Mostly for pedagogical reasons, we are now going to abandon x as our favorite indeterminate, giving preference to the indeterminate z in $\mathbb{R}[[z]]$. This may help us remember that the indeterminate is not an unknown real number or a variable real number.

This short section consists of only 1 example. It will be easy to generalize this example only if you completely understand the example. You are urged to reread this important section until you have mastered its content.

We consider only the question, How many nonnegative integer solutions (e_1, e_2, e_3, e_4) are there to the system

$$e_1 + e_2 + e_3 + e_4 = r$$

such that $e_1 \leq 4$, e_2 is odd and less than 9, and both e_3 and e_4 are even and each is less than 10?

We could do this problem by PIE but it would be an absolute bear. There are just so many conditions that complicate things. We now illustrate a technique that allows algebra to do a lot of our thinking and complicated counting for us. We change the problem above to one that is readily done by computing machines.

For any given nonnegative integer r, let a_r be the answer to our problem. Then a_r is the coefficient of z^r in the polynomial $g(z)$ in indeterminate z where $g(z)$ is

$$(z^0 + z^1 + z^2 + z^3 + z^4)(z^1 + z^3 + z^5 + z^7)(z^0 + z^2 + z^4 + z^6 + z^8)^2.$$

Why? Because to get the terms in the expansion of $g(z)$ that involve z^r, we take z^{e_1} from the first factor, we take z^{e_2} from the second factor, and, counting the square as 2 factors, we take z^{e_3} from the third factor and we take z^{e_4} from the fourth factor. So

$$z^{e_1} z^{e_2} z^{e_3} z^{e_4} = z^{e_1 + e_2 + e_3 + e_4} = z^r.$$

Hence,

$$e_1 + e_2 + e_3 + e_4 = r$$

with exactly the same restrictions as given by the system above, namely, $e_1 \leq 4$, e_2 is odd and less than 9, and both e_3 and e_4 are even and each is less than 10. We have changed our original question to one concerning the computation of a coefficient in a formal power series. The latter is the kind of thing computer algebra systems are extremely good at. This is a very clever trick!

We call the expression $g(z)$ above a "generating function" only for historical reasons. Rather than think of $g(z)$ as a mapping or function, think of $g(z)$ simply as an element of $\mathbb{R}[[z]]$. As a mental crutch, here and usually in the first step in writing down a generating function, we resist the conventions of writing z^0 as 1 and of writing z^1 as z. It is the exponents that have our full attention; and if it is hard to to write z and think 1, it is even harder to write 1 and think 0.

§26. Clotheslines

For formal power series f and g in $\mathbb{R}[[z]]$, we allow ourselves to write f/g iff there exists a power series h in $\mathbb{R}[[z]]$ such that $hg = f$, even though g may not itself have a reciprocal (multiplicative inverse). So, if e^z is defined to be $\sum_{i=0}^{\infty} z^i/i!$ in $\mathbb{R}[[z]]$, then $(e^z - 1)/z$ and $1/(1-z)$ make sense, even though $5z/0$ and $1/z$ are just nonsense.

The Definition. For a given sequence $\{a_r | r = 0, 1, 2, 3, \dots\}$, with the a_r in $\mathbb{R}$, the **generating function** for the sequence is the power series $g(z)$ in indeterminate z such that

$$g(z) = \sum_{r=0}^{\infty} a_r z^r.$$

Remember that in spite of the historical use of "function" in the definition above, we should think of z as an indeterminate and not as a real variable. Thus, our generating functions are elements of $\mathbb{R}[[z]]$. Only custom has prevented us from calling these power series "generating series" or simply "generators," rather than "generating functions."

We might think of a generating function as a clothesline onto which we have pinned the elements of a given sequence of real numbers. Of course, generating functions are more than a showcase for sequences. As we will see, it is the addition and multiplication of generating functions that make them so valuable. Although generating functions themselves have infinitely many terms, the definition of addition and multiplication given at the beginning of this chapter do not involve infinite sums of real numbers, as, for example, the coefficient of z^n in $f(z)g(z)$ is a finite sum of $n + 1$ terms in $\mathbb{R}$.

We observe that it is easy to show that in $\mathbb{R}[[z]]$ the element $\sum_{i=0}^{\infty} a_i z^i$ has an inverse iff $a_0 \neq 0$, that is, given $\sum_{i=0}^{\infty} a_i z^i$, there exists $\sum_{j=0}^{\infty} b_j z^j$

such that

$$1 = \left[\sum_{i=0}^{\infty} a_i z^i \right] \left[\sum_{j=0}^{\infty} b_j z^j \right]$$

iff $a_0 \neq 0$. This comes from looking at the product of the generating functions on the right side of the equation above and observing that we must have

$$1 = a_0 b_0$$
$$0 = a_0 b_1 + a_1 b_0$$
$$0 = a_0 b_2 + a_1 b_1 + a_2 b_0$$
$$0 = a_0 b_3 + a_1 b_2 + a_2 b_1 + a_3 b_0$$
$$0 = a_0 b_4 + a_1 b_3 + a_2 b_2 + a_3 b_1 + a_4 b_0$$

$$\vdots$$

The b_i are successively defined, in order, by the lines above when and only when $a_0 \neq 0$.

We now consider some illustrations of the definition of a generating function. For fixed positive integer n, what is the generating function for the sequence $\{a_r\}$ with $a_r = \binom{n}{r}$? With the usual convention that $\binom{n}{r} = 0$ for $r > n \geq 0$, this one is easy. After thinking about whether the following makes sense and checking our proof of the Binomial Theorem (or more realistically, not even thinking about it at all), we replace a by z and replace b by 1 in the our statement of the Binomial Theorem $(a+b)^n = \sum_{r=0}^{n} \binom{n}{r} a^r b^{n-r}$ to get

$$(1 + z)^n = \sum_{r=0}^{n} \binom{n}{r} z^r = \sum_{r=0}^{\infty} \binom{n}{r} z^r.$$

So, the generating function for $\{a_r\}$ with $a_r = \binom{n}{r}$ is $(1 + z)^n$.

Reinterpreting the formula for the sum of a finite geometric series, we can answer the following question. For a given fixed positive integer n, what is the generating function for $\{a_r\}$ with $a_r = 1$ for $0 \leq r \leq n$ and with $a_r = 0$ for $r > n$? The answer is $g(z)$ where

$$g(z) = 1 + z + z^2 + z^3 + z^4 + \cdots + z^n = \frac{1 - z^{n+1}}{1 - z}.$$

Likewise, reinterpreting the formula for the sum of an infinite geometric series, we see that the generating function for $\{a_r\}$ with $a_r = 1$ for $0 \leq r$ is $g(z)$ where

$$g(z) = 1 + z + z^2 + z^3 + z^4 + \cdots + z^r + \cdots = \frac{1}{1 - z}.$$

Each of the last 2 generating functions can be proved very easily by checking the equality simply by multiplying both sides of the equation by $1 - z$, without any recourse to real geometric series. We will see that the last is a special case, when we answer the question, What is the coefficient of z^r in the expansion of $(1 + z + z^2 + z^3 + z^4 + \cdots)^n$? Before that, however, we will generalize the last in a different way. If V is any element in $\mathbb{R}[[z]]$ except 1, then by expanding the product of $1 - V$ and $1 + V + V^2 + V^3 + \cdots$ we see that

$$1 + V + V^2 + V^3 + V^4 + \cdots + V^r + \cdots = \frac{1}{1 - V}.$$

A particularly useful case of this equation is when $V = z^k$ for positive integer k, which produces

$$1 + z^k + z^{2k} + z^{3k} + z^{4k} + \cdots + z^{rk} + \cdots = \frac{1}{1 - z^k}.$$

The given product $(1 + z + z^2 + z^3 + z^4 + \cdots)^n$ can be written as

$$[z^0 + z^1 + z^2 + z^3 + z^4 + \cdots][z^0 + z^1 + z^2 + z^3 + z^4 + \cdots] \cdots [z^0 + z^1 + z^2 + z^3 + z^4 + \cdots],$$

where there are n factors, each enclosed in a pair of brackets. As usual, we have written 1 as z^0 and z as z^1 to help us concentrate on the exponents. For an easy illustration, we find the coefficient of z^2 in this product. We see that z^2 can be obtained in the n ways of multiplying together z^2 from 1 factor within brackets and the z^0 from each of the remaining $n - 1$ factors within brackets; otherwise, z^2 can be obtained only from multiplying together z^1 from 2 of the factors within brackets and z^0 from each of the remaining $n - 2$ factors within brackets. Since, for this second way of obtaining z^2 in the product, we can choose the 2 factors that contribute the z^1 in $\binom{n}{2}$ ways, then the coefficient of z^2 must be $n + \binom{n}{2}$, which is $\left\langle \binom{n}{2} \right\rangle$. In the general case, we see that z^r is obtained in the product by choosing z^{e_1} from the first factor, choosing z^{e_2} from the second factor, choosing z^{e_3} from the third factor, and so on, until choosing z^{e_n} from the last factor under the condition that $e_1 + e_2 + e_3 + \cdots + e_n = r$. Therefore, the number we are seeking is the number of nonnegative integer solutions to the equation $e_1 + e_2 + e_3 + \cdots + e_n = r$, which we know to be $\left\langle \binom{n}{r} \right\rangle$. Also, since $(1 + z + z^2 + z^3 + z^4 + \cdots)(1 - z) = 1$, we can write $(1 + z + z^2 + z^3 + z^4 + \cdots)^n = \left[\frac{1}{1-z} \right]^n$. The generating function for the sequence $\{a_r\}$ with $a_r = \left\langle \binom{n}{r} \right\rangle$ for a given n is $\left(\frac{1}{1-z} \right)^n$.

Observation.

$$\sum_{r=0}^{\infty} \left\langle \binom{n}{r} \right\rangle z^r = (1 + z + z^2 + z^3 + z^4 + \cdots)^n = \left(\frac{1}{1-z} \right)^n = (1 - z)^{-n}.$$

We collect the 3 basic formulas to have them in 1 place for easy reference.

Three Basic Formulas.

$$(1+z)^n = \sum_{r=0}^{n} \binom{n}{r} z^r.$$

$$1 + z + z^2 + z^3 + z^4 + \cdots + z^n = \frac{1 - z^{n+1}}{1 - z}.$$

$$\sum_{r=0}^{\infty} \left\langle \binom{n}{r} \right\rangle z^r = (1 + z + z^2 + z^3 + z^4 + \cdots)^n = \left(\frac{1}{1-z} \right)^n.$$

We will use these formulas when we briefly look at the problem of computing, by hand, coefficients of a given generating function. Much more important is the modeling of problems. That is, modeling hard problems by reducing them to the algebraic problem of finding the coefficient of some power of z in a product of generating functions. The basic idea is to let the algebra consider all the cases for us. There is the paradox that in finding a_k for a particular value of k essentially requires finding an expression giving a_r for all nonnegative values of r. Nevertheless, we are grateful for a method that reduces a hard problem to an algorithm. We will consider some more examples.

How many ways are there to select 25 balls from unlimited supplies of red, white, and blue balls provided that at least 3 red balls are selected and at most 4 white balls are selected? The generating function for selecting red balls is $z^3 + z^4 + z^5 + \cdots$. The generating function for selecting white balls is $z^0 + z^1 + z^2 + z^3 + z^4$. The generating function for selecting blue balls is $z^0 + z^1 + z^2 + \cdots$. Thus, the generating function $g(z)$ for selecting r balls is

$$(z^3 + z^4 + z^5 + \cdots)(z^0 + z^1 + z^2 + z^3 + z^4)(z^0 + z^1 + z^2 + z^3 + \cdots).$$

Hence, the answer to our question is the coefficient of z^{25} in $g(z)$.

How many ways are there to select r balls from 3 red balls, 4 white balls, and 3 blue balls? In this case, the generating function for selecting red balls is $z^0 + z^1 + z^2 + z^3$. The generating function for selecting white balls is $z^0 + z^1 + z^2 + z^3 + z^4$. The generating function for selecting blue balls is $z^0 + z^1 + z^2 + z^3$. Thus, the generating function for selecting r balls is

$$(z^0 + z^1 + z^2 + z^3)(z^0 + z^1 + z^2 + z^3 + z^4)(z^0 + z^1 + z^2 + z^3).$$

Our answer is the coefficient of of z^r in $(1+z+z^2+z^3)^2(1+z+z^2+z^3+z^4)$.

How many positive integer solutions of $x_1 + x_2 + x_3 + x_4 = 25$ are such that $x_i \leq 4$ for $i = 1$ and $i = 2$? The generating function for the number of solutions is $z^1 + z^2 + z^3 + z^4$ for x_1 and for x_2 and is $z^1 + z^2 + z^3 + z^4 + \cdots$ for x_3 and for x_4. So the answer to our question is the coefficient of z^{25} in $g(z)$ where

$$
\begin{aligned}
g(z) &= (z^1 + z^2 + z^3 + z^4)^2 (z^1 + z^2 + z^3 + z^4 + \cdots)^2 \\
&= (z + z^2 + z^3 + z^4)^2 (z + z^2 + z^3 + z^4 + \cdots)^2 \\
&= z^2 (1 + z + z^2 + z^3)^2 z^2 (1 + z + z^2 + z^3 + \cdots)^2 \\
&= z^4 \left(\frac{1 - z^4}{1 - z} \right)^2 \left(\frac{1}{1 - z} \right)^2 \\
&= z^4 (1 - z^4)^2 \left(\frac{1}{1 - z} \right)^4 .
\end{aligned}
$$

How many ways are there to distribute 5 red balls, 6 white balls, and 7 blue balls to Peter and Paul so that each gets 9 balls and at least 1 ball of each color? We can model the problem by considering what balls Peter gets, since Paul gets whatever Peter does not. The generating functions for distributing, in turn, red, white, and blue balls to Peter are $z^1 + z^2 + z^3 + z^4$, $z^1 + z^2 + z^3 + z^4 + z^5$, and $z^1 + z^2 + z^3 + z^4 + z^5 + z^6$. (Note, for example, that Peter can not get all 5 red balls since Paul must get at least 1.) So our answer is the coefficient of z^9 in

$$(z^1 + z^2 + z^3 + z^4)(z^1 + z^2 + z^3 + z^4 + z^5)(z^1 + z^2 + z^3 + z^4 + z^5 + z^6).$$

Another way to model the same problem is to first give each of Peter and Paul a ball of each color, and then pose the question, How many ways are there to select 6 balls (for Peter) from 3 red balls, 4 white balls, and 5 blue balls? This time the answer is the coefficient of z^6 in

$$(z^0 + z^1 + z^2 + z^3)(z^0 + z^1 + z^2 + z^3 + z^4)(z^0 + z^1 + z^2 + z^3 + z^4 + z^5),$$

or

$$(1 + z + z^2 + z^3)(1 + z + z^2 + z^3 + z^4)(1 + z + z^2 + z^3 + z^4 + z^5).$$

Homework.

If "$\sum_{r=0}^{\infty} a_r z^r$" is the answer, what is the question? Now, use a generating function to model each of the following problems.

1. How many ways are there to distribute r red balls to 6 persons such that each person gets k balls where $3 \leq k \leq 5$?

2. How many ways can r ice cream cones (single-dip) be ordered, with repetition, from 8 available flavors?

3. How many ways are there to put r red balls into n distinguishable boxes if at most 2 red balls are distributed to each of the first 2 boxes?

4. How many ways are there to put r red balls into n distinguishable boxes if at most 2 red balls are distributed between the first 2 boxes?

5. How many 10-combinations of the letters A,B,C,D,E are possible if A can be used any number of times, B must be used at least once, C can be used at most once, D is used exactly twice or not at all, and E is used an even number of times?

§27. Examples and Homework

How many ways are there to make r cents change from unlimited amounts of pennies, nickels, and dimes? The generating function for selecting the pennies is $z^0 + z^1 + z^2 + z^3 + z^4 + \cdots$. No problem there. Now, the generating function for selecting the nickels is $z^0 + z^5 + z^{10} + z^{15} + z^{20} + \cdots$ because we are calculating everything in cents; the z^{15} here means that we have selected 3 nickels and so 15 cents in change. It is instructive to think of this generating function for the nickels as $(z^5)^0 + (z^5)^1 + (z^5)^2 + (z^5)^3 + (z^5)^4 + \cdots$ because we can see the number of nickels as an exponent to z^5, where each nickel is 5 cents in change. Likewise the generating function for selecting the dimes is $(z^{10})^0 + (z^{10})^1 + (z^{10})^2 + (z^{10})^3 + (z^{10})^4 + \cdots$, or $z^0 + z^{10} + z^{20} + z^{30} + z^{40} + \cdots$. The answer to our question is the coefficient of z^r in the generating function $g(z)$ for making r cents in change with pennies nickels, and dimes, where

$$
\begin{aligned}
g(z) = {}& [z^0 + z^1 + z^2 + z^3 + \cdots] \\
& \times [(z^5)^0 + (z^5)^1 + (z^5)^2 + (z^5)^3 + \cdots] \\
& \qquad \times [(z^{10})^0 + (z^{10})^1 + (z^{10})^2 + (z^{10})^3 + \cdots] \\
= {}& (1 + z + z^2 + z^3 + \cdots) \\
& \times (1 + z^5 + z^{10} + z^{15} + z^{20} + \cdots) \\
& \qquad \times (1 + z^{10} + z^{20} + z^{30} + z^{40} + \cdots) \\
= {}& \frac{1}{1-z} \cdot \frac{1}{1-z^5} \cdot \frac{1}{1-z^{10}} .
\end{aligned}
$$

It should be easy to extend this result to answer the question, How many ways are there to make r cents change using American coins?

How many nonnegative integer solutions are there for the equation

$$
x_1 + 5x_2 + 10x_3 = r?
$$

As Yogi Berra said, "It's déjà vu all over again." In other words, this question has the same generating function, and therefore the same answer, as the previous question. To see this, interpret x_1 as the number of pennies, x_2 as the number of nickels, and x_3 as the number of dimes in the previous problem. It cannot be mentioned too often that translating one problem to another that is familiar is a most important technique in combinatorics.

What is the generating function for the sequence $\{a_r\}$ where a_r is the number of solutions in nonnegative integers to $x_1+x_2+x_3+x_4+x_5+x_6 = r$ where either all x_i are even or else all x_i are odd? Confirm that the answer is $g(z)$ where

$$g(z) = \left(\frac{1}{1-z^2}\right)^6 + \left(\frac{z}{1-z^2}\right)^6 = (1+z^6)\left(\frac{1}{1-z^2}\right)^6.$$

Find the generating function $g(z)$ for the sequence $\{a_r\}$ where a_r is the number of ways 2 indistinguishable dice can show a sum of r. Here, we have to calculate each coefficient individually. From Table 1 of The Back of the Book on page 184, we have the solution

$$g(z) = z^2 + z^3 + 2z^4 + 2z^5 + 3z^6 + 3z^7 + 3z^8 + 2z^9 + 2z^{10} + z^{11} + z^{12}.$$

Homework.

Use a generating function to model each of the first 7 problems (and, for the courageous, also #8).

1. How many ways can 10 balls be selected from a bag of 7 red balls, 8 white balls, and 7 blue balls?

2. How many ways can r coins be selected from unlimited supplies of nickels, dimes, and quarters?

3. How many ways are there to distribute n indistinguishable balls to 10 persons with each receiving 0, 5, or 9 balls?

4. How many nonnegative integer solutions are there to the equation $2x_1 + 3x_2 + 4x_3 = 66$?

5. How many ways are there to distribute n balls to 4 persons if red and blue balls are available such that each receives at least 2 balls of each color?

6. How many ways are there to distribute n balls to 5 persons if red and blue balls are available?

7. How many ways are there to obtain a sum of 35 when 10 distinguishable dice are rolled?

8. How many ways are there to obtain an even sum when 10 indistinguishable dice are rolled? (Hint: Let x_i be the number of dice showing the number i.)

Find the generating function $g(z)$ for the sequence $\{a_r\}$ where a_r is the number of integer solutions to

$$y_1 + y_2 + y_3 + y_4 = r \text{ such that } 0 \leq y_1 \leq y_2 \leq y_3 \leq y_4.$$

The solution requires some ingenuity. Let $y_1 = x_1, y_2 = y_1 + x_2, y_3 = y_2 + x_3,$ and $y_4 = y_3 + x_4$. Then

$$y_1 = x_1 \qquad = x_1,$$
$$y_2 = y_1 + x_2 = x_1 + x_2,$$
$$y_3 = y_2 + x_3 = x_1 + x_2 + x_3, \text{ and}$$
$$y_4 = y_3 + x_4 = x_1 + x_2 + x_3 + x_4.$$

It follows that

$$r = y_1 + y_2 + y_3 + y_4 = 4x_1 + 3x_2 + 2x_3 + x_4 \text{ with } x_i \geq 0.$$

Therefore we have

$$g(z) = \frac{1}{1-z} \cdot \frac{1}{1-z^2} \cdot \frac{1}{1-z^3} \cdot \frac{1}{1-z^4}.$$

Thinking of black and white, we let b and w be indeterminates and extend the idea of a generating function to an array $\{a_{r,s}\}$ where is $a_{r,s}$ is the coefficient of $b^r w^s$ in $g(b, w)$. To be formal, we would say $g(b, w)$ is in $\mathbb{R}[[b, w]]$, which is defined by $\mathbb{R}[[b, w]] = (\mathbb{R}[[b]])[[w]]$, and show that $wb = bw$; we shall be informal and accept the "obvious." The purpose of these power series in 2 indeterminates is to have a nice place to hang the numbers $a_{r,s}$. We will have no qualms about using power series in several indeterminates.

For example, the generating function for the array $\{a_{r,s}\}$ where $a_{r,s}$ is the number of ways to distribute r black balls and s white balls to 6 persons is $g(b, w)$ where

$$g(b, w) = [(b^0 + b^1 + b^2 + b^3 + \cdots)(w^0 + w^1 + w^2 + w^3 + \cdots)]^6$$
$$= \left(\frac{1}{1-b}\right)^6 \left(\frac{1}{1-w}\right)^6.$$

Of course, we know, without using generating functions, that $a_{r,s} = \binom{6}{r}\binom{6}{s}$.

The generating function for the array $\{a_{r,s}\}$ where $a_{r,s}$ is the number of ways to "paint" all the vertices of a fixed square such that r vertices are black and s vertices are white and such that the vertices on a diagonal have the same color is $g(b,w)$ where $g(b,w) = (b^2 + w^2)^2$. This follows from the fact that the generating function for the pair of vertices on each of the 2 diagonals is $b^2 + w^2$ because these 2 vertices are either both black or else both white. Here, we have $g(b,w) = (b^2 + w^2)^2 = b^4 + 2b^2w^2 + w^4$. In total, there are exactly 4 ways to paint the vertices, as expected. The $1 + 2 + 1$ ways are 1 way all black, 2 ways with 2 black and 2 white, and 1 way all white.

More Homework.

Model each of the following 7 problems with a generating function.

1. How many ways are there to solve $2x_1 + 3x_2 + 4x_3 = 66$ with positive integers?

2. How many ways are there to obtain a sum of 40 when 10 distinguishable dice are rolled?

3. How many ways are there to distribute 20 red balls to 4 persons provided Jack and Jill together get no more than 5 balls and 1 of Peter and Paul gets an odd number while the other gets an even number?

4. How many ways are there to spend 25 dollars if each day for a week we spend 2 dollars for a red ball, a white ball, or a blue ball, spend 3 dollars for a black ball, or else spend 4 dollars for a green ball or an orange ball?

5. How many ways are there to distribute 20 of 32 red balls to 5 persons provided Adam, Eve, and Steve together get no more than 5 balls, while Lucky and Lucy together get at least 4 balls.

6. Argue that replacing the dice in a crap game by a pair of dice where 1 die has sides numbered $1, 2, 2, 3, 3, 4$ and the other die has sides numbered $1, 3, 4, 5, 6, 8$ should not change the game in any way because of the identity

$$(z^1 + 2z^2 + 2z^3 + z^4)(z^1 + z^3 + z^4 + z^5 + z^6 + z^8)$$
$$= (z^1 + z^2 + z^3 + z^4 + z^5 + z^6)^2.$$

7. How many ways can we paint the vertices of a fixed square such that each vertex is either black or white and there are the same number black as white? (Of course the answer is 6, but remember that we are after the model.)

Depending on interests, it is possible to skip the rest of this section and also all except the first 2 paragraphs of the next section.

How many ways are there to distribute r red balls into 4 distinguishable boxes such that the second box has more balls than the first box? Here, we want the number of nonnegative integer solutions to

$$x_1 + x_2 + x_3 + x_4 = r \text{ where } 0 \leq x_1 < x_2.$$

Let $x_2 = x_1 + 1 + x_0$. So $x_0 \geq 0$, and now we want the number of nonnegative integer solutions to $1 + x_0 + 2x_1 + x_3 + x_4 = r$. Our answer is the coefficient of z^r in $g(z)$ where

$$g(z) = z \cdot \frac{1}{1-z} \cdot \frac{1}{1-z^2} \cdot \frac{1}{1-z} \cdot \frac{1}{1-z} = \frac{z}{1-z^2} \left(\frac{1}{1-z}\right)^3.$$

The number $\Pi(r)$ of partitions of r is the same as the number of integer solutions e_i to the system

$$1e_1 + 2e_2 + 3e_3 + \cdots + re_r = r \text{ with } 0 \leq e_i.$$

The generating function for $\Pi(r)$ is seen to be $g(z)$ where

$$\begin{aligned}
g(z) = &[(z^1)^0 + (z^1)^1 + (z^1)^2 + (z^1)^3 + \cdots + (z^1)^n + \cdots] \\
&\times [(z^2)^0 + (z^2)^1 + (z^2)^2 + (z^2)^3 + \cdots + (z^2)^n + \cdots] \\
&\times [(z^3)^0 + (z^3)^1 + (z^3)^2 + (z^3)^3 + \cdots + (z^3)^n + \cdots] \\
&\times [(z^4)^0 + (z^4)^1 + (z^4)^2 + (z^4)^3 + \cdots + (z^4)^n + \cdots] \\
&\vdots \\
&\times [(z^k)^0 + (z^k)^1 + (z^k)^2 + (z^k)^3 + \cdots + (z^k)^n + \cdots] \\
&\vdots \\
= &\frac{1}{(1-z)(1-z^2)(1-z^3)(1-z^4)\cdots(1-z^k)\cdots}.
\end{aligned}$$

In evaluating a coefficient for a particular value of r, we would let $(1 - z^r)$ be the last factor in the denominator in the quotient above when turning to the computer for a calculation. Why?

In our last example, we will use the following, which is easily proved by multiplying both sides of the equation in the conclusion by $1 - z$.

Observation.

If $g(z) = \sum_{i=0}^{\infty} a_i z^i$, then $\sum_{r=0}^{\infty} \left(\sum_{k=0}^{r} a_k\right) z^r = \frac{g(z)}{1-z}$.

We began this section by finding the number of nonnegative integer solutions to the equation $x_1 + 5x_2 + 10x_3 = r$. We found the generating function here to be $g(z)$ where $g(z) = \frac{1}{1-z} \cdot \frac{1}{1-z^5} \cdot \frac{1}{1-z^{10}}$. Taking $g(z) = \sum_{i=0}^{\infty} a_i z^i$, we end the section by finding the number of nonnegative integer solutions for the system

$$x_1 + 5x_2 + 10x_3 \le r \text{ with } x_i \ge 0.$$

Let $h(z)$ denote the generating function for this problem. Then

$$h(z) = \sum_{r=0}^{\infty} \left(\sum_{k=0}^{r} a_k \right) z^r = \frac{g(z)}{1-z} = \frac{1}{(1-z)^2(1-z^5)(1-z^{10})},$$

where the first equality may require some thought, the second follows from the observation above, and the last follows from our previous result. Our answer is the coefficient of z^r in $h(z)$.

§28. Computations

We will do 2 different types of computation here. The second is hand calculation of answers to a couple problems that are like the ones we have done above. The first is an algebraic computation that involves e^{az}, for real number a, and is defined to be an abbreviation for $\sum_{i=0}^{\infty} \frac{(az)^i}{i!}$ in $\mathbb{R}[[z]]$. We want to prove the law of exponents: $e^{az}e^{bz} = e^{(a+b)z}$ for real numbers a and b.

Now, in general,

$$\left[\sum_{i=0}^{\infty} a_i \frac{z^i}{i!} \right] \left[\sum_{j=0}^{\infty} b_j \frac{z^j}{j!} \right] = \sum_{r=0}^{\infty} \left[\sum_{i+j=r} a_i \frac{z^i}{i!} b_j \frac{z^j}{j!} \right]$$

$$= \sum_{r=0}^{\infty} \left[\sum_{i+j=r} a_i b_j \frac{(i+j)!}{i!j!} \frac{z^{i+j}}{(i+j)!} \right]$$

$$= \sum_{r=0}^{\infty} \left[\sum_{k=0}^{r} \binom{r}{k} a_k b_{r-k} \right] \frac{z^r}{r!}$$

and so, in particular,

$$e^{az}e^{bz} = \left[\sum_{i=0}^{\infty} \frac{(az)^i}{i!} \right] \left[\sum_{j=0}^{\infty} \frac{(bz)^j}{j!} \right] = \sum_{r=0}^{\infty} \left[\sum_{k=0}^{r} \binom{r}{k} a^k b^{r-k} \right] \frac{z^r}{r!}$$

$$= \sum_{r=0}^{\infty} [(a+b)^r] \frac{(z)^r}{r!} = \sum_{r=0}^{\infty} \frac{((a+b)z)^r}{r!}$$

$$= e^{(a+b)z},$$

as desired.

Now for something completely different. How many ways are there to express r as a sum of distinct positive integers, with the terms in increasing order? The generating function for this problem is $g(z)$ where

$$g(z) = (1+z)(1+z^2)(1+z^3)(1+z^4)\cdots(1+z^k)\cdots.$$

So we want the coefficient of z^r in $g(z)$. Now, for a particular value of r, to be practical we drop all the factors in $g(z)$ that have exponent greater than r since these will contribute nothing to our answer. Even so, the calculation is horrendous. This is exactly the kind of thing best left to computers. Certainly, with computers available, knowing how to model problems with generating functions is much more valuable than knowing how to calculate by hand the coefficients of given power series. However, some familiarity of this process will do no harm, as long as we keep away from calculations that are too long. A couple examples follow.

How many ways can 40 voters cast their 40 votes for 5 candidates such that no candidate gets more than 10 votes?

First, note that this problem is equivalent to asking, How many integer solutions are there to the system $e_1 + e_2 + e_3 + e_4 + e_5 = 40$ with $0 \le e_i \le 10$? Thinking of the variables e_i as exponents, we see that we are asking for the coefficient a_{40} of z^{40} in the product

$$[1 + z + z^2 + z^3 + \cdots + z^{10}]^5 =$$
$$[z^0 + z^1 + z^2 + \cdots + z^{10}][z^0 + z^1 + z^2 + \cdots + z^{10}]\cdots[z^0 + z^1 + z^2 + \cdots + z^{10}]$$
$$= [(1 - z^{11})/(1 - z)]^5 = [1 - z^{11}]^5 \left[\frac{1}{1 - z}\right]^5 = \sum_{r=0}^{\infty} a_r z^r.$$

We know that with

$$\sum_{i=0}^{\infty} b_i z^i = (1 - z^{11})^5 = \sum_{i=0}^{5} (-1)^i \binom{5}{i} z^{11i} \text{ and}$$
$$\sum_{j=0}^{\infty} c_j z^j = \left[\frac{1}{1 - z}\right]^5 = \sum_{j=0}^{\infty} \left\langle \begin{matrix} 5 \\ j \end{matrix} \right\rangle z^j,$$

the answer to our question is a_{40} where

$$a_{40} = b_0 c_{40} + b_{11} c_{29} + b_{22} c_{18} + b_{33} c_7$$
$$= \binom{5}{0}\left\langle \begin{matrix} 5 \\ 40 \end{matrix} \right\rangle - \binom{5}{1}\left\langle \begin{matrix} 5 \\ 29 \end{matrix} \right\rangle \binom{5}{2}\left\langle \begin{matrix} 5 \\ 18 \end{matrix} \right\rangle - \binom{5}{3}\left\langle \begin{matrix} 5 \\ 7 \end{matrix} \right\rangle.$$

Think about the form of this answer. Interpret each of the terms in the answer.

Finally, we consider the following 5 equivalent problems, where we suppose that $2n \leq r \leq 4n$.

1. How many ways are there to put r indistinguishable balls into n distinguishable boxes with 2, 3, or 4 balls in each box?

2. How many integer solutions are there to the system

$$e_1 + e_2 + \cdots + e_n = r \text{ with } 2 \leq e_i \leq 4?$$

3. What is the coefficient of z^r in the product $[z^2 + z^3 + z^4]^n$?

4. What is the coefficient of z^r in the product $z^{2n}[1 - z^3]^n \left[\frac{1}{1-z}\right]^n$?

5. What is the coefficient of z^r in the product

$$[z^{2n}] \left[\sum_{i=0}^{n} (-1)^i \binom{n}{i} z^{3i} \right] \left[\sum_{j=0}^{\infty} \left\langle \begin{matrix} n \\ j \end{matrix} \right\rangle z^j \right] ?$$

To get the terms in the product that contribute to the coefficient of z^r in the product of the 3 series within brackets, we necessarily take z^{2n} from the first series, then a term of the form $(-1)^i \binom{n}{i} z^{3i}$ from the second series, and finally the term from the third series so that the sum of all the exponents of z is r. Hence, the last term must be of the form $c_j z^j$, where $j = r - 2n - 3i$ and $c_j = \left\langle \begin{matrix} n \\ r-2n-3i \end{matrix} \right\rangle$. With it understood that $\left\langle \begin{matrix} n \\ r-2n-3i \end{matrix} \right\rangle = 0$ if $3i > r - 2n$, we see that the answer to each of our 5 questions is

$$\sum_{i=0}^{n} (-1)^i \binom{n}{i} \left\langle \begin{matrix} n \\ r-2n-3i \end{matrix} \right\rangle, \quad \text{or} \quad \sum_{i=0}^{n} (-1)^i \binom{n}{i} \binom{r-n-3i}{n-1}.$$

Again, we should think about the form of the answer and be able to interpret each of the terms in this answer. It might be clear that generating functions are handy for PIE problems, especially when there are many complicated conditions.

§29. The Greater Mississippi Problem; A Look at Exponential Generating Functions

The generating functions that we have studied are also called *ordinary generating functions* when other forms are in evidence. The one of these other forms that is of interest here is called an exponential generating function. After the formal definition below, we shall give 2 problems that are paradigms for the application of exponential generating functions.

For the sequence $\{a_r | r = 0, 1, 2, 3, \ldots\}$ with $a_r \in \mathbb{R}$, the **exponential generating function** for the sequence is the power series $h(z)$ in indeterminate z where

$$h(z) = a_0 + a_1 z + \frac{a_2}{2!}z^2 + \frac{a_3}{3!}z^3 + \frac{a_4}{4!}z^4 + \cdots + \frac{a_r}{r!}z^r + \cdots$$

$$= a_0 + a_1 z + a_2 \frac{z^2}{2!} + a_3 \frac{z^3}{3!} + a_4 \frac{z^4}{4!} + \cdots + a_r \frac{z^r}{r!} + \cdots = \sum_{r=0}^{\infty} a_r \frac{z^r}{r!}.$$

Since $(1+z)^n = \sum_{r=0}^{n} \binom{n}{r} z^r = \sum_{r=0}^{\infty} \binom{n}{r} z^r = \sum_{r=0}^{\infty} \frac{n!}{(n-r)!} \cdot \frac{z^r}{r!}$, then the exponential generating function (**egf**) for the sequence $\{a_r\}$ where $a_r = n!/(n-r)!$ is $(1+z)^n$. Thus, for a given fixed n, the ordinary generating function for the sequence $\left\{ \binom{n}{r} \right\}$ that counts the number of r-combinations of n things is exactly the same as the egf for the sequence $\left\{ \frac{n!}{(n-r)!} \right\}$ that counts the number of r-permutations of n things.

Paradigm #1. How many 5-letter words can be formed from the 11 letters of MISSISSIPPI ?

Since we can choose the M exactly 0 times or 1 time,
the egf for selecting M's is $\frac{z^0}{0!} + \frac{z^1}{1!}$.
Since we can choose an I exactly 0, 1, 2, 3, or 4 times,
the egf for selecting I's is $\frac{z^0}{0!} + \frac{z^1}{1!} + \frac{z^2}{2!} + \frac{z^3}{3!} + \frac{z^4}{4!}$.
Since we can choose an S exactly 0, 1, 2, 3, or 4 times,
the egf for selecting S's is $\frac{z^0}{0!} + \frac{z^1}{1!} + \frac{z^2}{2!} + \frac{z^3}{3!} + \frac{z^4}{4!}$.
Since we can choose a P exactly 0, 1, or 2 times,
the egf for selecting P's is $\frac{z^0}{0!} + \frac{z^1}{1!} + \frac{z^2}{2!}$.
Now, as with the ordinary generating functions, we consider the product $h(z)$ of these individual egf, even though this does not look very promising at first. So $h(z)$ is

$$\left[\frac{z^0}{0!} + \frac{z^1}{1!} \right] \left[\frac{z^0}{0!} + \frac{z^1}{1!} + \frac{z^2}{2!} + \frac{z^3}{3!} + \frac{z^4}{4!} \right] \left[\frac{z^0}{0!} + \frac{z^1}{1!} + \frac{z^2}{2!} + \frac{z^3}{3!} + \frac{z^4}{4!} \right] \left[\frac{z^0}{0!} + \frac{z^1}{1!} + \frac{z^2}{2!} \right].$$

Suppose we take the term $\frac{z^{e_i}}{e_i!}$ from the ith factor above to form the product $\frac{z^{e_1}}{e_1!} \cdot \frac{z^{e_2}}{e_2!} \cdot \frac{z^{e_3}}{e_3!} \cdot \frac{z^{e_4}}{e_4!}$. This corresponds to counting all the words that have exactly e_1 M's, e_2 I's, e_3 S's, and e_4 P's. Certainly we already know how many such words there are. This is our basic Mississippi problem and we get the result $\frac{(e_1+e_2+e_3+e_4)!}{e_1!e_2!e_3!e_4!}$. Now watch and be amazed as this number magically appears as the coefficient of $\frac{z^{e_1+e_2+e_3+e_4}}{(e_1+e_2+e_3+e_4)!}$ in the calculation

$$\frac{z^{e_1}}{e_1!} \cdot \frac{z^{e_2}}{e_2!} \cdot \frac{z^{e_3}}{e_3!} \cdot \frac{z^{e_4}}{e_4!} = \frac{1}{e_1!e_2!e_3!e_4!} \cdot \frac{z^{e_1+e_2+e_3+e_4}}{1}$$

$$= \frac{(e_1 + e_2 + e_3 + e_4)!}{e_1!e_2!e_3!e_4!} \cdot \frac{z^{e_1+e_2+e_3+e_4}}{(e_1 + e_2 + e_3 + e_4)!}.$$

For our particular problem, we see that we want the sum of all coefficients of $\frac{z^{e_1+e_2+e_3+e_4}}{(e_1+e_2+e_3+e_4)!}$ in the product $h(z)$ where $e_1 + e_2 + e_3 + e_4 = 5$. That is, the coefficient of $z^5/5!$ in $h(z)$ is the number of 5-letter words that can be formed from the letters of MISSISSIPPI. Is it clear that the number of r-letter words that can be formed from the letters in MISSISSIPPI is the coefficient of $z^r/r!$ in $h(z)$? We observe that the egf are precisely constructed so that we can multiply the individual egf together to get the egf for the entire process. The multiplication of series does all the algebraic thinking and calculating for us. Again we have reduced a complicated problem to an algorithm of computing the proper coefficient in a product of series. A computer algebra system tells us that the desired coefficient that answers the question posed in the first paradigm is 550. We can hope that we are not asked to hand calculate the solution to the question, How many words of at least 1 letter can be formed from the 11 letters of MISSISSIPPI?

We can follow the paradigm to use egf to solve problems modeled on arranging r letters from an "alphabet" of n symbols, in particular where there are limitations on the availability of the different letters.

There are some special egf that are very useful: With unlimited repetition allowed in selecting a letter or any object for an arrangement, the egf for that object is $\frac{z^0}{0!} + \frac{z^1}{1!} + \frac{z^2}{2!} + \frac{z^3}{3!} + \frac{z^4}{4!} + \cdots$, which is e^z.
The egf for selecting at least 1 of an object is $e^z - 1$.
The egf for selecting at least 2 of an object is $e^z - 1 - z$.
The egf for selecting at least 3 of an object is $e^z - 1 - z - \frac{z^2}{2!}$.
The egf for selecting at least 4 of an object is $e^z - 1 - z - \frac{z^2}{2!} - \frac{z^3}{3!}$; etc.
The egf for selecting an even number of some object is $\frac{e^z+e^{-z}}{2}$ since

$$\frac{z^0}{0!} + \frac{z^2}{2!} + \frac{z^4}{4!} + \frac{z^6}{6!} + \frac{z^8}{8!} + \cdots = \frac{e^z + e^{-z}}{2}.$$

The egf for selecting an odd number of some object is $\frac{e^z-e^{-z}}{2}$ since

$$\frac{z^1}{1!} + \frac{z^3}{3!} + \frac{z^5}{5!} + \frac{z^7}{7!} + \frac{z^9}{9!} + \cdots = \frac{e^z - e^{-z}}{2}.$$

Some examples using these identities follow.

How many ways are there to form sequences of r letters from an alphabet of 5 letters?
The egf here is $(e^z)^5$. The coefficient of $z^r/r!$ in e^{5z} is 5^r. There are 5^r ways, which, of course, is no surprise.

How many ways are there to form sequences of r letters from an alphabet of 5 letters if each letter must be used at least once?
This time the egf is $(e^z - 1)^5$. Since, by the Binomial Theorem,

$$(e^z - 1)^5 = e^{5z} - 5e^{4z} + 10e^{3z} - 10e^{2z} + 5e^z - 1.$$

the answer to the question is $5^r - 5 \cdot 4^r + 10 \cdot 3^r - 10 \cdot 2^r + 5$, which we could have calculated by PIE.

How many ways are there to form n-digit ternary sequences, which are words of n letters from the alphabet $\{0, 1, 2\}$, such that the sequence contains an odd number of 0's and an even number of 1's? Our solution is the coefficient of $z^n/n!$ in the egf

$$\frac{e^z - e^{-z}}{2} \cdot \frac{e^z + e^{-z}}{2} \cdot e^z.$$

Since the egf is $\frac{e^{3z} - e^{-z}}{4}$, then the answer to this nontrivial problem is

$$\frac{3^n - (-1)^n}{4}.$$

Three Exercises.

Model the following with egf.

1. How many r-letter words containing only vowels are there such that A is used at least once, E is used an even number of times, I is used an odd number of times, and each of O and U is used exactly twice or not at all?

2. How many r-letter words containing only vowels are there such that no vowel appears exactly once?

3. How many ways are there to pile red, white, blue, and green poker chips in a stack r high such that the stack has an even number of red chips and at least 1 white chip?

Paradigm #2. How many ways are there to put 8 distinguishable balls into 3 distinguishable boxes such that no box has more that 4 balls?

Using Figure 3.1, we see that there is a one-to-one correspondence between putting 8 distinguishable balls, numbered from 1 to 8, into 3 distinguishable boxes, labeled A, B, C, and forming 8-letter words from the 3-letter alphabet $\{A, B, C\}$. This result easily generalizes to the following.

Observation. There is a one-to-one correspondence between the placements of r distinguishable balls into n distinguishable boxes and r-letter words using an n-letter alphabet.

The solution to Paradigm #2 is then the same as the solution to the problem, How many 8-letter words can be formed from 4 A's, 4 B's, and 4 C's? Our answer is the coefficient of $z^8/8!$ in $\left(\frac{z^0}{0!} + \frac{z^1}{1!} + \frac{z^2}{2!} + \frac{z^3}{3!} + \frac{z^4}{4!} \right)^3$.

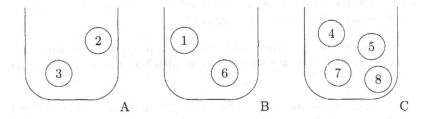

BAACBCC

FIGURE 3.1. A Useful Correspondence.

Exponential generating functions provide the ideal model for putting distinguishable balls into distinguishable boxes, especially so when there are complicated restrictions on the number of balls that can go into the particular boxes. The conditions on the boxes correspond to the conditions on the letters. In passing, we also note that if computers are useful for calculating with ordinary generating functions, they are even more useful with egf.

How many ways are there to distribute 5 distinguishable balls into 4 distinguishable boxes such that the first box has at most 1 ball, each of the second and third boxes has at most 4 balls, and the fourth box has at most 2 balls?

Labeling the 4 boxes M, I, S, P, in turn, and considering the correspondence given by Figure 3.1, we see that we are back to Paradigm #1.

How many ways are there to distribute r distinguishable balls into 5 distinguishable boxes if the first box has an even number of balls, the second box has an odd number of balls, the third box has at most 2 balls, the fourth box has at least 1 ball, and there is no restriction on the number of balls in the fifth box?

Since the egf is $h(z)$ where

$$h(z) = \frac{e^z + e^{-z}}{2} \cdot \frac{e^z - e^{-z}}{2} \cdot \left[\frac{z^0}{0!} + \frac{z^1}{1!} + \frac{z^2}{2!} \right] \cdot [e^z - 1] \cdot e^z,$$

our answer is the coefficient of $z^r/r!$ in $h(z)$.

Four Exercises.

Model the following with egf.

1. How many ways are there to assign each of r persons to 1 of 6 distinguishable rooms with between 2 and 4 assigned to each room?

2. How many ways can r distinguishable balls be put into 6 distinguishable boxes with at least 1 box empty?

3. How many ways are there to distribute r distinguishable balls into 5 distinguishable boxes such that the fifth box has a positive even number of balls?

4. How many ways are there to distribute 25 distinguishable balls into 5 distinguishable boxes if the first box has at least 2 balls, the second box has at most 3 balls, the third box has has an odd number of balls, there is no restriction on the number of balls in the fourth box, but the fifth box contains between 7 and 11 balls?

Five Review Exercises.

Model the following problems with generating functions, ordinary or exponential.

1. How many 5-letter words can be made from the 12 letters AAA BB C D EE FFF?

2. How many ways are there to distribute 26 of 34 distinguishable balls to 5 persons if Lucy gets at most 4 balls?

3. How many ways are there to distribute 35 orange balls to 6 persons if Adam and Eve together get at least 7 balls while Lucky gets at least 4 balls?

4. How many ways can 25 distinguishable balls be put in 5 distinguishable boxes if each box receives an odd number of balls?

5. How many ways are there to solve the system consisting of the equation $x_1 + 2x_2 + 3x_3 + x_4 + x_5 = 28$ and the inequality $x_1 + x_2 < 6$ in positive integers?

§30.   Comprehensive Exams

A Warm-up Exercise.

1. How many 4-letter words consisting only of A's, B's, C's, and D's have the first letter not A, the second letter not B, the third letter not C, and the fourth letter not D?

2. How many arrangements of $1, 2, 3, 4$ are there such that 1 is not in position 3, 2 is not in position 1, 3 is not in position 2, and 4 is not in position 4?

3. How many ways are there to distribute 14 distinguishable balls to 8 persons if the first 3 together get 5 balls?

4. How many ways are there to distribute 14 red balls to 8 persons if the first 3 together get 5 balls?

5. How many ways are there to distribute 14 balls to 8 persons if red and blue balls are available and the first 3 together get 5 balls?

Two Core Exams.

Core Exam #1.

Please read "¿" as "How many ways are there to". Do NOT simplify your answers. Questions #1–19, worth 4% each; questions #20–22, worth 10% each; questions #23–25, worth 8% each. Total 130%.

1. ¿ pick 2 books, not both on the same subject, from 5 algebra books, 7 geometry books, and 4 calculus books, where the books are distinguishable? **2.** ¿ select pieces of fruit from 5 apples and 7 oranges if at least 1 piece of fruit is chosen? **3.** ¿ seat 5 men and 9 women at a round table with no 2 men next to each other? **4.** ¿ form 5-letter words with the letters in alphabetical order? **5.** ¿ partition 18 persons into study groups of 5, 5, 4, and 4? **6.** ¿ arrange the letters in MISSISSIPPI with at least 2 I's adjacent? **7.** ¿ permute the first 20 positive integers? **8.** ¿ arrange the 13 letters of ALBANYNEWYORK with the consonants in alphabetical order? **9.** ¿ have 9 persons, including Peter and Paul, sit in a row with Peter and Paul not sitting next to each other? **10.** ¿ form 10-letter words from the alphabet without repeating any letter? **11.** ¿ fly 7 red flags and 20 blue flags on 10 distinguishable flag poles? **12.** ¿ fly 7 red flags, 20 blue flags, and the flags of 12 different nations on 10 distinguishable flag poles if each flagpole must have at least 1 flag? **13.** ¿ pick 28 letters from 14 A's, 14 B's and 14 C's? **14.** ¿ have 9 dice fall? **15.** ¿ pass out some (including all and none) of 12 oranges and 13 distinguishable books to 8 girls? **16.** ¿ seat 11 persons at a round table? **17.** ¿ put 12 pennies, 12 dimes, 12 quarters, and 12 Anthony dollars into 5 distinguishable boxes? **18.** ¿ put 25 indistinguishable balls in 9 distinguishable boxes if each box must have an odd number of balls? **19.** ¿ select 5 of the first 20 positive integers such that the difference of any 2 is greater than 1? **20.** ¿ arrange the letters of ALBANYNEWYORK with both A's preceding both N's or both Y's preceding the K? **21.** ¿ arrange the letters of ALBANYNEWYORK with no double letter? **22.** ¿ put 20 distinguishable balls in 4 distinguishable boxes with no box empty? **23.** ¿ arrange n letters selected from SUNYATALBANY? **24.** ¿ have 10 distinguishable dice show a sum of 30? **25.** ¿ distribute 20 red balls to 4 persons provided Jack and Jill together get no more than 5 balls and 1 of Peter and Paul gets an odd number while the other gets an even number?

Core Exam #2.

Please read "¿" as "How many ways are there to". Do NOT simplify your answers. Questions #1–20, worth 3% each; questions #21–24, worth 10% each; questions #25–27, worth 6% each. Total 118%.

1. ¿ arrange the consonants of the alphabet? **2.** ¿ form 5-digit positive integers having the last digit a repeat? **3.** ¿ form 3-digit positive integers having exactly one 8? **4.** ¿ form 5-element subsets of $\{1, 2, 3, \ldots, 30\}$ such that the largest is 20? **5.** ¿ form 5-element subsets of $\{1, 2, 3, \ldots, 25\}$ such that the largest is greater than 20? **6.** ¿ select 5 of the letters of the alphabet, having no consecutive pair? **7.** ¿ arrange the letters in MISSISSIPPI with the M preceding all S's? **8.** ¿ arrange the letters in MISSISSIPPI with no adjacent S's? **9.** ¿ arrange the letters in MISSISSIPPI with adjacent P's? **10** ¿ arrange the letters in MISSISSIPPI? **11.** ¿ put 20 distinguishable balls into 7 distinguishable boxes? **12.** ¿ pick at least 2 pieces of fruit from 7 apples and 11 oranges? **13.** ¿ put 30 red balls into 7 distinguishable boxes with exactly 2 boxes empty? **14.** ¿ put 30 red balls into 7 distinguishable boxes? **15.** ¿ distribute some (including none and all) of 7 apples and 11 oranges to 8 children? **16.** ¿ seat 6 men and 8 women at a round table with no 2 men next to each other? **17.** ¿ invite at least 1 of 12 friends to dinner? **18.** ¿ select 6 of 14 knights seated at a round table (to release the enchanted princess) such that no 2 sitting adjacent are selected. **19.** ¿ fly 15 indistinguishable flags and 17 different, distinguishable flags from 12 flagpoles? **20.** ¿ fly 15 indistinguishable flags and 17 different, distinguishable flags from 12 flagpoles if each flagpole has at least 1 flag? **21.** ¿ distribute 7 red balls, 8 white balls, and 9 blue balls to 2 persons if each receives 12 balls? **22.** ¿ distribute 5 pennies, 5 nickels, 5 dimes, and 5 quarters to 4 persons so that each person receives at least 1 coin? **23.** ¿ arrange the letters of COMBINATORICS so that the arrangement contains 2 adjacent letters that are the same? **24.** ¿ select 10 coins from 9 pennies, 6 nickels, 4 dimes, and 3 quarters? **25.** ¿ solve the equation $2x + 3y + 4z = 66$ with positive integers? **26.** ¿ obtain a sum of 40 when 10 distinguishable dice are rolled? **27.** ¿ distribute 30 red balls to 4 persons provided Jack and Jill together get no more than 10 balls, Peter gets an odd number, and Paul gets an even number?

4
Groups

§31. Symmetry Groups

Let ρ be a rotation of 90° about the origin O in the plane. (Remember that convention dictates that ρ is then a counterclockwise rotation of 90°.) So, ρ^2 denotes the rotation of 180° about O, and ρ^3 denotes the rotation of 270° about O, that is, ρ^3 denotes 3 successive rotations of 90° about O. Let σ be the reflection in the X–axis and $\iota = \sigma^2$. Then ι is the identity mapping on the plane, sending each point of the plane to itself, and σ maps the point (x, y) to the point $(x, -y)$. We write $\sigma((x, y)) = (x, -y)$. Mappings, like these, that fix distance are called **isometries** and are multiplied under **composition**. This means, for example, that $\sigma\rho$, which is usually read "sigma rho," can be more completely read "sigma following rho" and is defined by the formula

$$\sigma\rho(P) = \sigma(\rho(P)) \text{ for any point } P.$$

Hence, to find the image of point P under the mapping $\sigma\rho^2$ we first see where P goes under ρ^2, say to point Q, and then see where Q goes under σ. Therefore, if we have $\sigma(Q) = R$, then we have $\sigma\rho^2(P) = \sigma(\rho^2(P)) = \sigma(Q) = R$. (You may very well agree with the author and think that this convention for composition is backwards because we read English from left to right, while composition is performed from right to left. We bow to this convention, since it is the the convention always used in calculus, where $gf(x) = g(f(x))$ for functions f and g and real number x. However, we

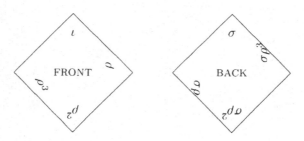

FIGURE 4.1. Marking a Square.

will usually ignore those that think we should write $\sigma\rho$ as "$\sigma \circ \rho$" to avoid confusion with something that we are not even going to mention.)

A **symmetry** for a set of points is an isometry that sends the set to itself. Considering the symmetries of a square in the *standard position* of having $(1,0)$ as a vertex and the center of the square at O, we see that there are exactly 8 symmetries. (Any of the 4 vertices can go to $(1,0)$; next, either of the 2 vertices adjacent to the first can go to $(0,1)$; and this then automatically determines the position of the other vertices.) Since the following 8 symmetries of the square are distinct, they must, in fact, be exactly the 8 symmetries of the square:

$$\iota, \ \rho, \ \rho^2, \ \rho^3, \ \text{and} \ \sigma, \ \sigma\rho, \ \sigma\rho^2, \ \sigma\rho^3.$$

Is it clear that the product of 2 symmetries of a set of points is a symmetry of that set, as is the inverse of a symmetry?

Exercise 1. Rip off an end of a sheet of regular paper to form a square. With the square oriented as in Figure 4.1, mark an ι at the top. Rotate the square $90°$ and label the top ρ. Rotate another $90°$ and label the top ρ^2. Finally, another $90°$ and label the top ρ^3. Now, with the square in its original position with the ι at the top, reflect the square in its horizontal axis. This represents the isometry σ that is defined by $\sigma((x,y)) = (x,-y)$ (and is physically achieved in 3-space by rotating the square about its horizontal axis). Doing the reflections after the rotations again, we end up with the square marked as in Figure 4.1. It should be very clear that $\sigma\rho \neq \rho\sigma$. With the aid of this marked square, actually carry out the physical motions necessary to determine the Cayley table (multiplication table under composition) for the symmetry group D_4 of the square. That is, fill in a copy of Table 4.1. Remember that the entry for $(\sigma\rho)(\sigma)$ goes in the row headed $\sigma\rho$ and column headed σ.

Exercise 2. Show that composition of mappings is always associative: give the reasons for the following steps in proving that $\gamma(\beta\alpha) = (\gamma\beta)\alpha$, where

D_4	ι	ρ	ρ^2	ρ^3	σ	$\sigma\rho$	$\sigma\rho^2$	$\sigma\rho^3$
ι								
ρ								
ρ^2								
ρ^3								
σ								
$\sigma\rho$								
$\sigma\rho^2$								
$\sigma\rho^3$								

TABLE 4.1. Cayley Table for D_4.

α, β, γ are any mappings for which the compositions are defined. For any point P, we have:

$$(\gamma(\beta\alpha))(P) = \gamma((\beta\alpha)(P))$$
$$= \gamma(\beta(\alpha(P)))$$
$$= (\gamma\beta)(\alpha(P))$$
$$= ((\gamma\beta)\alpha)(P).$$

In general, C_n denotes the **cyclic group** of order n generated by ρ where ρ is the rotation $\rho_{O,360/n}$ of $(360/n)°$ about the origin. Thus, the elements of C_n are $\iota, \rho, \rho^2, \ldots, \rho^{n-1}$, with multiplication given by $\rho^i\rho^j = \rho^{i+j}$ and $\rho^n = \iota$. Let σ be the reflection in the X-axis. The **dihedral group** D_n, which is the symmetry group of a regular n-gon when $n > 2$, has the $2n$ elements:

$$\iota, \rho, \rho^2, \ldots, \rho^{n-1}, \text{ and } \sigma, \sigma\rho, \sigma\rho^2, \ldots, \sigma\rho^{n-1}.$$

The group multiplication table for D_n is actually very easy to compute once we know that $\rho^k\sigma = \sigma\rho^{-k} = \sigma\rho^{n-k}$. (It was Leonardo da Vinci who first observed that any finite symmetry group in the plane is (conjugate to) either a C_n or a D_n.)

In applications later, we will be content to "describe" the 8 elements of D_4 in a more robust manner as:

1 identity,
2 rotations of $\pm 90°$,
1 rotation of $180°$,
2 reflections in diagonals,
2 reflections in perpendicular bisectors of sides.

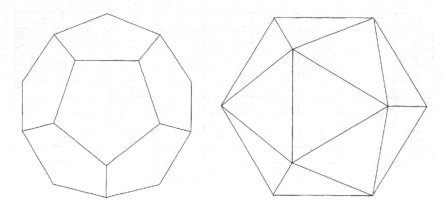

FIGURE 4.2. Dodecahedron and Icosahedron.

Later we will be precise about what we mean by a robust manner; for now, use the description above as a model and you will not go far wrong.

Exercise 3. The equations $24 = 8 \times 3 = 12 \times 2 = 6 \times 4$ might suggest that there are 3 ways to show that there are 24 rotation symmetries of a cube. Describe in a robust manner these symmetries of a cube. (Obviously, this exercise is best done with a cube in hand. We will be using the cube so much that it will pay to actually have a cube available. In a pinch, you can use a die or even use a cube that you have made from paper and tape. You should become very familiar with the rotation symmetries of the cube.) We will not be concerned with any symmetries that cannot be physically achieved in 3-space. For example, although the reflection in the plane containing a pair of opposite edges of the cube is an isometry of the cube, this motion is impossible in 3-space with a rigid cube.

Exercise 4. Describe in a robust manner the rotation symmetries of a regular dodecahedron and a regular icosahedron. (The dodecahedron has 12 sides that are congruent regular pentagons. The icosahedron has 20 sides that are congruent equilateral triangles. These solids have the same symmetries because joining the centers of adjacent faces of one produces the edges of one of the other.) See Figure 4.2.

The group of rotation symmetries of the regular dodecahedron and the regular icosahedron is called the **icosahedral group**. The group of the rotation symmetries of the cube is called the **octahedral group**. This is because the cube and the regular octahedron have the same rotation symmetries. Joining the centers of adjacent faces of one produces the edges of one of the other. Each of the 8 faces of the regular octahedron is an equilateral triangle.

§32.   Legendre's Theorem

Mathematicians have the peculiar propensity to use everyday words for technical mathematical purposes. The list of technical words adopted by mathematicians could start off with *rational, irrational, real, imaginary, complex, map, series, graph, field, ring,* and *group*. We do not need to define these here, except for the last of these words. Mathematicians studying various systems realized that they were duplicating their efforts and that the proof over here in one system was pretty much the same as the proof over there in another system. They began to abstract what was common to many of these systems in order to form a general theory. The most useful basic system involves a set and 1 binary operation (there is another word: an *operation* in mathematics has nothing to do with hospitals or the military). By a **binary operation** on set S, we mean a mapping of all the ordered pairs of elements from S into some set. For example, addition $+$ on the set of integers. We put the operation symbol between the first and second arguments of a binary operation. So, we do not write "$+(3,4) = 7$," but rather "$3 + 4 = 7$." After all, people used "$3 + 4$" long before there was the abstract idea of a binary operation. In general, we call the binary operation *multiplication*, although in some instances, depending on the notation, we call the operation *addition*. If that sounds indecisive, it is. The properties in the following list are those that are deemed most important for a set S together with a binary operation $\odot$ on S.

Closure Property. When we combine 2 elements of S we should get an element of S. In this case, we say that S is **closed** under the binary operation. So, if α and β are in S and if S is closed under $\odot$, then $\alpha \odot \beta$ is an element of S.

Associative Property. Most nice binary systems satisfy the **associative law**: if α, β, and γ are elements in S, then

$$\gamma \odot (\beta \odot \alpha) = (\gamma \odot \beta) \odot \alpha.$$

Actually, life gets very difficult studying systems that do not have this property.

Identity Property. There is be a unique element ι in S such that for each α in S we have

$$\iota \odot \alpha = \alpha \odot \iota = \alpha.$$

This unique element is called an **identity**. The most familiar identities are 0 for addition on the real numbers, 1 for multiplication on the nonzero real numbers, and the identity mapping on any particular set. If S is a set of mappings, we will usually denote the identity mapping on S by ι. (Note that the meaning ι changes as we change

S. Most of us have the bad habit of talking about *the* identity, when we really mean the relevant identity of the moment. Somehow, we get used to 1 being the identity one minute and 0 the identity the next.)

Inverse Property. If α is in S, then there is a unique element α^{-1} in S such that

$$\alpha^{-1} \odot \alpha = \alpha \odot \alpha^{-1} = \iota,$$

where ι is the identity. This unique element α^{-1} is called the **inverse** of element α.

Observe that we have not mentioned the popular **commutative law**: $\beta \odot \alpha = \alpha \odot \beta$, for all α and β in S. Although the associative law is pretty much a necessity, the "law" in "commutative law" is frequently flaunted. If a binary operation does satisfy the commutative law, then the operation is said to be **commutative** or **abelian**.

The triple $(S, \odot, \iota)$ such that S is a set containing the element ι and $\odot$ is a binary operation on S satisfying all the 4 properties above is a **group**. The properties can be weakened somewhat, but we are more interested in what a group *is* than how most economically to define a group. Since "group" is a technical word, mathematicians shouldn't use the word to refer to a general collection or set. (That's "shouldn't" and not "don't.") Just to make matters worse, unfortunately we often refer to the set of a group as the group itself when it is clear what the binary operation on the set is. For example, we feel safe in mentioning "the group of integers," because we know that everyone will understand that the binary operation here is garden variety addition and that the identity is 0. Understanding the jargon and using the jargon is part of mastering any field.

Besides those groups that come from the arithmetic of real numbers, we have the symmetry groups C_n and D_n from the previous section as examples of groups where the binary operation is composition and the group identity is the identity map on a set of elements. Those who encountered formal power series in the previous chapter will see that the set of series with nonzero constant term form a group under Cayley multiplication $\otimes$ with 1 as the identity.

Exercise 1. Prove the **cancellation laws** for a group. That is, prove that if α, β, and γ are in group G, then

$$\gamma \odot \alpha = \gamma \odot \beta \text{ implies } \alpha = \beta,$$

and

$$\alpha \odot \gamma = \beta \odot \gamma \text{ implies } \alpha = \beta.$$

Exercise 2. Show: In any group, the inverse of a product is the product of the inverses in reverse order.

Another place where mathematicians intentionally allow themselves to be sloppy is in a statement such as "H is a subgroup of G." We suppose that H is a subset of G. If $\odot$ is a binary operation on G and if $H \neq G$, then $\odot$ is not really a binary relation on H. Admittedly, the difference is subtle. There is a *restriction* $\boxdot$ of $\odot$ to H defined by $\alpha \boxdot \beta = \alpha \odot \beta$ for all α and β in H. However, nothing is gained by being too precise. We don't think many of us are upset in reading "3 + 4" at not being told if that " + " means addition on the set of integers or addition on the set of real numbers. So, we tacitly agree that to say H is a **subgroup** of group G when we formally mean, given group $(G, \odot, \iota)$ and a subset H of G, that $(H, \boxdot, \iota)$ is a group with $\boxdot$ defined by $\alpha \boxdot \beta = \alpha \odot \beta$ for all α and β in H. Of course, we then immediately drop the $\boxdot$ and $\odot$ and use · for the operations in both G and H. (Paradoxically, it is only the best students that are annoyed by the relentless abuse of notation mathematicians allow themselves.) For an easy example of a subgroup, the set of even integers forms a group under addition of the group of all integers under addition. Enough of this diddling; let's move on. (Somebody has to do the diddling, of course, but it does not have to be us.)

If H is a subgroup of group G and g is an element of G, then we define the **coset** gH to be $\{gh \mid h \in H\}$. In other words, we take the set of all the elements of H and multiply each on the left by g; the resulting set is gH. You probably did not even notice that we have fallen into another convention and used juxtaposition to denote our binary operation. Thus, we write "$\alpha\beta$" in place of "$\alpha \cdot \beta$." We have suppressed the symbol denoting the binary operation, but it will not be missed. (It must be mentioned that you might recognize from some other life that this definition of a coset is a left coset, might know that there are right cosets, and might know that much of the first course in abstract algebra has to do with when left cosets and right cosets are equal and the consequences when this is the case. We can get by very well without all that.)

In general, $|S|$ denotes the the number of elements in set S and is called the **size** of S or the **order** of S. The **order** of group G is the order of its set of elements and is denoted by $|G|$. The **empty set** is the set having no elements and is denoted by $\emptyset$. So, $|\emptyset| = 0$.

Lemma. Suppose that H is a subgroup of group G, that G has identity e, and that x and y are in G. Then

1. x is in xH. (Every element of G is in at least 1 coset of H.)

2. $hH = H$ for every h in H.

3. $xH = yH$ iff $y^{-1}x$ is in H.

4. Either $xH = yH$ or else $xH \cap yH = \emptyset$. (Every element of G is in at most 1 coset of H.)

5. $|xH| = |H|$ for every x in G.

Proof. (1) Since $e \in H$, then $xe \in xH$. Since e is the identity, then $xe = x$. So $x \in xH$.

(2) Suppose $h \in H$. Then $hH \subseteq H$ because H is closed. Also, if $x \in H$, then $h^{-1}x \in H$ and we have $x = h(h^{-1}x) \in hH$. So, $H \subseteq hH$. Hence, $hH = H$.

(3) Suppose $xH = yH$. Then there is an h in H such that $x = xe = yh$. So, $y^{-1}x = h \in H$. Conversely, suppose $y^{-1}x = h \in H$. Then $y^{-1}xH = H$ and so $xH = yH$.

(4) Suppose $z \in xH \cap yH$. Then there are h_1 and h_2 in H such that $z = xh_1 = yh_2$. So, $y^{-1}x = h_2h_1^{-1} \in H$ and, hence, $xH = yH$.

(5) This follows from the definition of a coset and the cancellation laws. ∎

If H is a subgroup of group G, then the number of different cosets H has is called the **index** of H in G and is denoted by $|G : H|$. By (1) and (4) above, we see that every element of G is in exactly 1 coset of H. From (5) it then follows that $|G : H| \cdot |H| = |G|$. This relation is usually stated as follows.

Lagrange's Theorem. The order of a subgroup divides the order of the group.

The hallmark of twentieth-century mathematics was the extensive blossoming of abstract mathematics. There is great economy in a proof in an abstract system such as a groups. We have just proved that the statement of Lagrange's Theorem holds not only for every group that has been conceived but also holds for any group that ever will be conceived. Now that is power.

§33.   Advanced Gozintas (Permutation Groups)

A **permutation** on a set is simply a bijection on that set, i.e., a one-to-one mapping of the set onto itself. In other words, a permutation on S is a one-to-one correspondence of S with itself. Often the word "permutation" is restricted to finite sets, as it will be here. Unless otherwise indicated, the set is usually supposed to be $\{1, 2, 3, \ldots, n\}$. Since all defined mappings are associative and a bijection automatically has an inverse, the permutations on any given set form a group. The group of permutations on $\{1, 2, 3, \ldots, n\}$

S_3	$\begin{pmatrix} 1 & 2 & 3 \\ 1 & 2 & 3 \end{pmatrix}$	$\begin{pmatrix} 1 & 2 & 3 \\ 2 & 3 & 1 \end{pmatrix}$	$\begin{pmatrix} 1 & 2 & 3 \\ 3 & 1 & 2 \end{pmatrix}$	$\begin{pmatrix} 1 & 2 & 3 \\ 1 & 3 & 2 \end{pmatrix}$	$\begin{pmatrix} 1 & 2 & 3 \\ 3 & 2 & 1 \end{pmatrix}$	$\begin{pmatrix} 1 & 2 & 3 \\ 2 & 1 & 3 \end{pmatrix}$
$\begin{pmatrix} 1 & 2 & 3 \\ 1 & 2 & 3 \end{pmatrix}$						
$\begin{pmatrix} 1 & 2 & 3 \\ 2 & 3 & 1 \end{pmatrix}$						
$\begin{pmatrix} 1 & 2 & 3 \\ 3 & 1 & 2 \end{pmatrix}$						
$\begin{pmatrix} 1 & 2 & 3 \\ 1 & 3 & 2 \end{pmatrix}$						
$\begin{pmatrix} 1 & 2 & 3 \\ 3 & 2 & 1 \end{pmatrix}$						
$\begin{pmatrix} 1 & 2 & 3 \\ 2 & 1 & 3 \end{pmatrix}$						

TABLE 4.2. Cayley Table for S_3.

is denoted as S_n and is called the **symmetric group** on n elements. Where x goes into $\alpha(x)$ under permutation α, we can write α as

$$\alpha = \begin{cases} 1 \to \alpha(1) \\ 2 \to \alpha(2) \\ \quad \vdots \\ n \to \alpha(n) \end{cases}$$

but this takes up too much room. It is much more convenient to use the following notation.

$$\alpha = \begin{pmatrix} 1 & 2 & 3 & \cdots & n \\ \alpha(1) & \alpha(2) & \alpha(3) & \cdots & \alpha(n) \end{pmatrix}.$$

It is evident that there are exactly $n!$ elements in S_n and that the inverse of any particular element in S_n is easily obtained by reading "up" instead of "down" in the array displayed above. Since permutations are mappings we shall follow the (stupid?) conventional left-handed notation for functions: $(\beta\alpha)(x) = \beta(\alpha(x))$. So, for example, we must read from right to left in the product on the left side of the equation to get the correct result on the right side of the equation

$$\begin{pmatrix} 1 & 2 & 3 \\ 1 & 3 & 2 \end{pmatrix} \begin{pmatrix} 1 & 2 & 3 \\ 2 & 3 & 1 \end{pmatrix} = \begin{pmatrix} 1 & 2 & 3 \\ 3 & 2 & 1 \end{pmatrix}.$$

Here, we have 1 goes into 2 under α, and 2 goes into 3 under β. Hence, 1 goes into 3 under the product $\beta\alpha$. Next, 2 goes into 3 under α, and 3 goes into 2 under β. Hence, 2 goes into 2 under the product $\beta\alpha$. Finally, 3 goes into 1 under α, and 1 goes into 1 under β. Hence, 3 goes into 1 under the product $\beta\alpha$.

Exercise 1. Compute the Cayley table for S_3 by filling in Table 4.2.

The notation for elements in S_n is still rather cumbersome. We shall streamline the notation with the **cycle notation**. The cycle notation is very much like what it sounds like. If α is in S_n, $\alpha(1) = 6$, $\alpha(6) = 2$, and $\alpha(2) = 1$, then the elements $1, 6, 2$ cycle in that order and we write this cycle as $(1, 6, 2)$. The cycles $(1, 6, 2)$, $(6, 2, 1)$, and $(2, 1, 6)$ are exactly the same and differ from $(1, 2, 6)$. Conventionally we start with the smallest integer in the cycle. We can write each element in S_n as a product of disjoint cycles. We start with 1 and capture the cycle containing 1, then start with the smallest integer not in any cycle yet obtained and determine the cycle that this element is in, and continue in this way until all the n integers have appeared. Since n is finite, this process will come to an end. For example, if

$$\alpha = \begin{pmatrix} 1 & 2 & 3 & 4 & 5 & 6 & 7 & 8 & 9 \\ 6 & 1 & 5 & 4 & 7 & 2 & 8 & 3 & 9 \end{pmatrix},$$

we get the cycles $(1,6,2)$, $(3,5,7,8)$, (4), and (9), in turn. Now we write

$$\alpha = (1, 6, 2)(3, 5, 7, 8)(4)(9).$$

Algebraists further streamline the notation by dropping all the cycles containing just 1 element and suppose that if an element is not mentioned it goes to itself; there is necessarily 1 exception to this: we write (1) for the identity permutation. For $n < 10$, we can also omit the commas without confusion. Thus, we now write $\alpha = (162)(3578)$.

If you are compulsive about sticking with left-handed notation you could write $\alpha = (3578)(162)$. The result is the same precisely because the cycles are disjoint. Of course, a product of cycles that are not disjoint is generally not commutative, as $(132) = (12)(13) \neq (13)(12) = (123)$.

A cycle $(e_1 e_2 \ldots e_k)$ is said to be of **length** k; a cycle of length 2 is called a **transposition**. An element in S_n is **even** or **odd** as the permutation is a product of an even or an odd number of transpositions. We will not show here that every permutation is either even or odd but not both.

Exercise 2. Rewrite, in cycle notation, the row and column headings from Table 4.2 for S_3 and recompute the Cayley table using the cycle notation.

Exercise 3. Show that in S_n with $n > 2$, an n-cycle is a product of $n - 1$ transpositions and that any product of 2 transpositions is a product of 3-cycles.

Exercise 4. List all the even elements of S_4; list all the odd elements of S_4. Show that the set of even elements of S_n forms a subgroup of S_n. This group is denoted by A_n and is called the **alternating group** on n elements. What is the order of A_n?

§34. Generators

Group G with identity e always has subgroups G and $\{e\}$. These subgroups are called **trivial** subgroups; a **proper** subgroup is a subgroup that is not trivial.

Suppose g is an element of group G. The product $ggg \cdots gg$, having r factors, is denoted by g^r. For positive integer r, we define g^{-r} to be $(g^{-1})^r$. So $g^r g^{-r} = g^0$, that is, $g^{-r} = (g^r)^{-1}$. For integer n, the element g^n is called a **power** of g. With $\mathbb{Z}$ denoting the set of integers, as usual, the set $\{g^n \mid n \in \mathbb{Z}\}$, consisting of all the powers of g, is easily seen to be a subgroup of G. For example, the set $\{2^n \mid n \in \mathbb{Z}\}$, which is $\{1, 2, \frac{1}{2}, 4, \frac{1}{4}, 8, \frac{1}{8}, \dots\}$, forms a subgroup of the positive rationals under multiplication. If there is a smallest positive integer r such that g^r is the identity element of G, then the set $\{g, g^2, \dots, g^r\}$ forms a finite subgroup of G.

If all the elements of a group G are powers of some element g in G, then we say G is **cyclic** and call g a **generator** of G. We abbreviate $\{g^n \mid n \in \mathbb{Z}\}$ to $\langle g \rangle$, which is read "the (cyclic) subgroup generated by g." So, if element h is in group H, then $\langle h \rangle$ is a cyclic subgroup of H. Thus $\langle 2 \rangle$ is a cyclic subgroup of the positive rationals under multiplication. The set of integers under addition is a cyclic group having each of $+1$ and -1 as a generator. If g is a generator of group G, then the inverse of g is necessarily also a generator of G. The integers mod 12 under addition, i.e., "clock arithmetic," form a cyclic group where each of 1, 5, 7, and 11 is a generator of the group. Let's check below that 5 is a generator.

First, we note that the notation for the group binary operation is addition rather than multiplication and so we have ng in place of g^n in the above, where ng is the sum $g + g + \cdots + g$ with n terms. Thus, with additive notation, $\langle g \rangle = \{ng \mid n \in \mathbb{Z}\}$. Under "clock arithmetic," we "add" 2 elements by doing the usual addition but then add or subtract 12 repeatedly until we get to an integer (inclusive) between 0 and 11. The group properties are quickly proved. (Here, 0 is the identity. The additive inverse of 5 is 7 since $5 + (12 - 5) = 0$. This group is denoted by $\mathbb{Z}_{12}$. Of course, there is nothing special about 12, and we have the analogous groups $\mathbb{Z}_n$ for positive integer n, which we read as "integers mod n." We want to check that 5 is a generator in $\mathbb{Z}_{12}$. So we need look at $\{5n \mid n \in \mathbb{Z}\}$, which is $\{5, 10, 3, 8, 1, 6, 11, 4, 9, 2, 7, 0\}$. Therefore, 5 generates all the elements of $\mathbb{Z}_{12}$. So, the group $\mathbb{Z}_{12}$, under addition, has 5 as a generator. Note, for example, that 9 is not a generator of $\mathbb{Z}_{12}$ since $\{9n \mid n \in \mathbb{Z}\}$ is $\{9, 6, 3, 0\}$, which is a proper subgroup of $\mathbb{Z}_{12}$. Since $\langle 9 \rangle \neq G$ then 9 is not a generator of G.

The **order** of element g in group G is the order of $\langle g \rangle$. (Note the 2 meanings of the 1 word.) As a consequence of Lagrange's Theorem, we have the following corollary.

Corollary. The order of an element in a group divides the order of the group.

The converse of the corollary states that if d divides the order of group G then there is an element of order d in the group. This is immediately seen to be false since not every group is cyclic. In fact, every proper subgroup of S_3 is cyclic but S_3 is not cyclic. That A_4 has no element of order 6 provides another counterexample. Does A_4 have any subgroup of order 6? Does the converse of Lagrange's Theorem hold?

The converse of Lagrange's Theorem does not hold. Assume A_4 has a subgroup K of order 6. (Certainly, 6 divides 12.) Let x be any element of order 3 in A_4. Since K has 6 elements in A_4, then at most 2 of the cosets K, xK, and x^2K are distinct. However, the equality of any 2 of the cosets implies that x is in K. Hence, K contains all 8 elements of order 3, a contradiction to $|K| = 6$. The Cayley table for A_4 is given in Table 4.3.

Exercise. List all the proper subgroups of A_4. Next, take 1 proper subgroup of each possible order and calculate all the cosets for that subgroup. (For example, there are 3 subgroups of order 2. Take only 1 of these, say $H = \{(1), (12)(34)\}$, and calculate all the cosets of H. Do the same for $H = \langle (123) \rangle$ and for the subgroup of order 4.)

The notation above can be generalized. If $X = \{g_1, g_2, \dots\}$ and each element of group G is a product of a finite number of powers of the g_i, where powers of an element of X may occur more than once in a product, then we say G is **generated** by X and write $G = \langle g_1, g_2, \dots \rangle$. For example, D_n is generated by a rotation ρ of order n and a reflection σ of order 2. The equations $D_n = \langle \rho, \sigma \rangle$ with $\rho^n = \sigma^2 = \iota$ and $\sigma\rho = \rho^{-1}\sigma$ completely describe the dihedral group D_n for any positive integer n. For another example, consider $\langle a, b \rangle$ where a is of order 6, b is of order 4, $b^2 = a^3$, and $ba = a^{-1}b$. This group of order 12 has no noncyclic proper subgroups but is not cyclic. We can express this group as a permutation group on twelve symbols by taking $a = (123456)(abcdef)$ and $b = (1f4c)(2e5b)(3d6a)$.

A permutation that is a cycle of length k has order k. A product of disjoint (!) cycles has order that is the least common multiple of the lengths of the cycles. For example, (12345) has order 5, and $(1234)(56)(789)$ has order 12.

A permutation in S_n has **cycle type** $z_{l_1}^{k_1} z_{l_2}^{k_2} z_{l_3}^{k_3} \cdots$ if the disjoint cycle decomposition of the permutation has, for $1 \leq i$, exactly k_i cycles of length l_i. The z_j's are indeterminates, onto which we hang the exponents, and we omit all factors having exponent 0. We suppose that $1 \leq l_1 < l_2 < \cdots$ and, for our purposes, suggest that we resist the usual convention of omitting to write 1 as an exponent. Already the bad habit of omitting cycles of length 1 in writing the elements of S_n can cause us to forget them here. A few examples should make it all clear. In S_9, permutation (1) has cycle type

A_4	(12)(34)	(13)(24)	(14)(23)	(123)	(132)	(124)	(142)	(134)	(143)	(234)	(243)	(1)
(12)(34)	(1)	(14)(23)	(13)(24)	(243)	(143)	(234)	(134)	(142)	(132)	(124)	(123)	(12)(34)
(13)(24)	(14)(23)	(1)	(12)(34)	(142)	(234)	(143)	(123)	(243)	(124)	(132)	(134)	(13)(24)
(14)(23)	(13)(24)	(12)(34)	(1)	(134)	(124)	(132)	(243)	(123)	(234)	(143)	(142)	(14)(23)
(123)	(134)	(243)	(142)	(132)	(1)	(13)(24)	(143)	(234)	(14)(23)	(12)(34)	(124)	(123)
(132)	(234)	(124)	(143)	(1)	(123)	(243)	(14)(23)	(12)(34)	(142)	(134)	(13)(24)	(132)
(124)	(143)	(132)	(234)	(14)(23)	(134)	(142)	(1)	(13)(24)	(243)	(123)	(12)(34)	(124)
(142)	(243)	(134)	(123)	(234)	(13)(24)	(1)	(124)	(132)	(12)(34)	(14)(23)	(143)	(142)
(134)	(123)	(142)	(243)	(124)	(14)(23)	(12)(34)	(234)	(143)	(1)	(13)(24)	(132)	(134)
(143)	(124)	(234)	(132)	(12)(34)	(243)	(123)	(13)(24)	(1)	(134)	(142)	(14)(23)	(143)
(234)	(132)	(143)	(124)	(13)(24)	(142)	(134)	(12)(34)	(14)(23)	(123)	(243)	(1)	(234)
(243)	(142)	(123)	(134)	(143)	(12)(34)	(14)(23)	(132)	(124)	(13)(24)	(1)	(234)	(243)
(1)	(12)(34)	(13)(24)	(14)(23)	(123)	(132)	(124)	(142)	(134)	(143)	(234)	(243)	(1)

TABLE 4.3. Cayley Table for A_4.

z_1^9, permutation $(12)(34)(567)(89)$ has cycle type $z_2^3 z_3^1$, and permutation $(23)(5678)$ has cycle type $z_1^3 z_2^1 z_4^1$. Group A_4 has 1 permutation of cycle type z_1^4, has 3 of cycle type z_2^2, and 8 of cycle type $z_1^1 z_3^1$. Further, S_4 has 6 additional permutations of cycle type z_4^1 and 6 of cycle type $z_1^2 z_2^1$. You have no doubt noticed that, in the notation at the beginning of this paragraph, we can use the equation $\sum_{i=1}^s k_i l_i = n$ as a check on the cycle type. Although it is probably difficult to imagine now, the cycle type of a permutation will turn out to be extremely useful.

§35.   Cyclic Groups

Cyclic groups are very special and we will briefly look at them.

Theorem. A cyclic group is abelian. Any subgroup of a cyclic group is cyclic.

Proof. Suppose G is a cyclic group generated by element g. So $G = \langle g \rangle$ for some element g in G and the identity element of G is g^0. Commutativity of G is obvious, following from the commutativity of the integers under addition:

$$g^r g^s = g^{r+s} = g^{s+r} = g^s g^r.$$

Let H be a subgroup of group G. If $H = \{g^0\}$, then $H = \langle g^0 \rangle$. Suppose $H \neq \langle g^0 \rangle$. Then there exists a smallest positive integer m such that g^m is in H. We shall show that $H = \langle g^m \rangle$. Suppose g^s is in H. Then there exist integers q and r such that $s = mq + r$ where $0 \leq r < m$. (We divide s by m to get quotient q with remainder r.) Thus, $g^r = g^{s-mq} = (g^s)(g^m)^{-q}$. Since the right-hand side of this equation is in H, then g^r is in H. However, by the minimality of m, we see that r must be 0 and $s = mq$. Thus, $g^s = (g^m)^q$. Therefore, every element of H is a power of g^m, as desired. ∎

If d is the largest positive integer that divides each of the positive integers n and s, then we say d is the **greatest common divisor** of n and s and we write $d = \gcd(n, s)$. If $\gcd(n, s) = 1$, then we say n and s are **relatively prime**. For example, $\gcd(4, 15) = 1$, $\gcd(4, 12) = 4$, and $\gcd(4, 30) = 2$. We introduce the standard notation $\phi(n)$ to denote the number of positive integers less than or equal to n that are relatively prime to n. The function ϕ, which is defined on the positive integers, is called **Euler's phi function**. Thus $\phi(1) = \phi(2) = 1$, $\phi(6) = 2$, and for prime p we have $\phi(p) = p - 1$ and $\phi(p^r) = p^r - p^{r-1}$. Evaluation of $\phi(n)$ for large values of n follows from the equation $\phi(ab) = \phi(a)\phi(b)$ if $\gcd(a, b) = 1$, which we will not prove here. We write "$d|n$" as an abbreviation for "d divides n," which means that for integers d and n there is an integer m such that $n = md$.

Theorem. If G is a cyclic group of order n with generator g, $1 \leq k \leq n$, and $\gcd(n, k) = m$, then $\langle g^k \rangle$ has order n/m.

Proof. Let $H = \langle g^k \rangle$. There exist integers s and t such that $n = sm$, $k = tm$, and $\gcd(s, t) = 1$. So $(g^k)^s = g^{ks} = g^{tms} = g^{tn} = (g^n)^t = g^0$. In other words, the order of g^k divides s and thus is less than or equal to s. So H contains the elements $g^k, g^{2k}, g^{3k}, \ldots, g^{sk}$. We need to show these s elements are distinct to prove the theorem. Assume that $g^{ik} = g^{jk}$ for some i and j such that $1 \le j < i \le s$. So $g^{(i-j)k} = g^0$. This means that n divides $(i - j)k$, which means that sm divides $(i - j)tm$, which means that s divides $(i - j)t$, which, since s and t are relatively prime, means that s divides $i - j$, which is impossible since $1 \le j < i \le s$. Therefore, H has order s where $s = n/m$. ∎

Since any subgroup of a cyclic group is cyclic, the theorem has some immediate corollaries. The first follows from the special case $m = 1$ in the theorem. The second is the special case $k = m = n/d$ in the theorem, so that $d = n/m$. The remaining 2 corollaries follow from the first 2 and Lagrange's Theorem

Corollary 1. If G is a cyclic group of order n with generator g, then $G = \langle g^k \rangle$ iff integers n and k are relatively prime.

Corollary 2. If G is a cyclic group of order n with generator g and $d|n$, then $\langle g^{n/d} \rangle$ has order d.

Corollary 3. Group C_n has $\phi(n)$ generators.

Corollary 4. Group C_n has a subgroup of order d iff $d|n$.

Theorem. Group C_n has exactly 1 subgroup of each order d such that d divides n, and these are the only subgroups of C_n.

Proof. We already know that every subgroup is cyclic and that $\langle \rho^{n/d} \rangle$ is a subgroup of order d. We need to show that this is the only subgroup of order d. So suppose H is a subgroup of order d. Then there exists s such that $H = \langle \rho^s \rangle$. From the previous theorem, we have $d = |H| = n/\gcd(n, s)$. Thus, $\gcd(n, s) = n/d$ and n/d must divide s. Hence, $\rho^s = (\rho^{n/d})^t$ for some integer t. Since the generator of H is in $\langle \rho^{n/d} \rangle$, then H is a subset of $\langle \rho^{n/d} \rangle$. However, since H and $\langle \rho^{n/d} \rangle$ have the same order, then the subgroups must be equal. ∎

This theorem has 2 corollaries that follow from the 3 statements: Each element of a group generates a cyclic subgroup C_d. All the elements in C_n that are of order d are in this unique cyclic subgroup of order d. A cyclic group of order d has $\phi(d)$ generators.

Corollary 1. If $d|n$, then C_n has $\phi(d)$ elements of order d.

Corollary 2. For positive integer n,

$$n = \sum_{d|n} \phi(d).$$

Homework. Let $G = C_{12} = \langle \rho \rangle$. So ρ is the rotation of $30°$ about the origin in the plane. Calculate the order of $\langle \rho^k \rangle$ for each k such that $1 \leq k \leq 12$. Use G to illustrate the last theorem and all of the last 6 corollaries. (If you need more exercises to understand the theorems and their corollaries, use $G = C_{36}$.)

§36. Equivalence and Isomorphism

This section could be called "The Same and More of the Same." Actually, these need some qualification such as "with respect to ... ". Of course, *same* brings to mind *equal*. Note that $a = b$ means that a and b are names for the same thing. If we give this some thought, we might conclude that mathematics is probably the only place where equality exists. In the physical world—at least at the normal scale—we see that 2 things cannot be equal because they are different and, even if indistinguishable, are necessarily in different places. We want to make precise the idea of things being equal with respect to some particular property.

To say $\sim$ is a **relation** on set S means only that $\sim$ is a subset of $S \times S$, the set of ordered pairs of elements from S. We write $a \sim b$ iff $(a, b) \in \sim$. We say the relation $\sim$ is an **equivalence relation** iff the following 3 properties hold: the *reflexive law*, the *symmetric law* and the *transitive law*. See Figure 4.3. These 3 properties go back to Aristotle's description of equality, the mother of all equivalence relations.

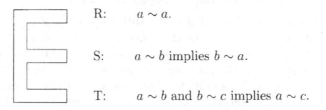

R: $a \sim a$.

S: $a \sim b$ implies $b \sim a$.

T: $a \sim b$ and $b \sim c$ implies $a \sim c$.

FIGURE 4.3. Laws for an Equivalence Relation.

We can think of a lot of examples from plane euclidean geometry. An equivalence relation on the lines is defined by $a \sim b$ iff $a \parallel b$. Segments under congruence, angles under congruence, and triangles under congruence provide further examples. (There is a very good reason to use the same word *congruence* here to describe the 3 different equivalence relations, but to explain would be too much of a digression. Although Euclid did use 1 word for equal and equivalent, there is no reason we should. It is, at best, old-fashioned to say that 2 triangles are equal when they are only congruent.)

For a given equivalence relation $\sim$, the **equivalence class** $\boxed{a}$ of a with $a \in S$ is $\{x \in S \mid x \sim a\}$.

Exercise 1. Prove that, for a given equivalence relation, equivalence classes $\boxed{a}$ and $\boxed{c}$ are identical or else disjoint. So, $\boxed{a} = \boxed{c}$ iff $a \sim c$.

Recall that a partition of a set S is a set of nonempty subsets of S such that each element of S is in exactly 1 of the subsets. By the exercise above, the equivalence classes of equivalence relation $\sim$ form a partition of S. Conversely, any partition of S defines the equivalence relation $\sim$ on S that is defined by $a \sim b$ iff a and b are in the same subset of the partition.

We will look at an another example by looking seriously at $\mathbb{Z}_{12}$ under "addition." (There is nothing special about 12. The sequel applies to any positive integer.) Rather than our notation $\sim$ for a general relation, here we will use the standard notation $\equiv$ and define $a \equiv b \pmod{12}$ iff 12 divides $a - b$. We read $a \equiv b \pmod{12}$ as "a is congruent to b mod 12," where "modulo" frequently replaces "mod." This is quickly seen to be an equivalence relation on the set $\mathbb{Z}$ of integers. Here, there are 11 equivalence classes. A couple examples are

$$\boxed{0} = \{\ldots, -36, -24, -12, 0, 12, 24, 36, 48, \ldots\}$$
$$= \{12z \mid z \in \mathbb{Z}\},$$
$$\boxed{1} = \{\ldots, -35, -23, -11, 1, 13, 25, 37, 49, \ldots\}$$
$$= \{12z + 1 \mid z \in \mathbb{Z}\},$$
$$\boxed{5} = \{\ldots, -31, -19, -7, 5, 17, 29, 41, 53, \ldots\}$$
$$= \{12z + 5 \mid z \in \mathbb{Z}\}.$$

We next define an addition $\oplus$ on the equivalence classes as follows

$$\boxed{a} \oplus \boxed{c} = \boxed{a+c}.$$

We are obligated to show that this definition is well defined. That is, we must argue that if a and a' are in the same equivalence class and if c and c' are in the same equivalence class, then $a + c$ and $a' + c'$ are in the same equivalence class. (Otherwise, "$\boxed{a} \oplus \boxed{c}$" doesn't make any sense. However, we omit the details.) Being totally lazy, of course, we write $a + c$ in place of $\boxed{a} \oplus \boxed{c}$. (This undoubtedly upsets your nonmathematical friends and neighbors who see you write $3 + 4 = 2$, not realizing that you know that the 3 is not really 3 but is the equivalence class $\boxed{3}$ in $\mathbb{Z}_5$ and that the plus sign indicates addition $\oplus$ on equivalence classes and is not addition on integers.)

If p is a prime, then we can define a multiplication on the equivalence classes of the "nonzero" elements of $\mathbb{Z}_p$ by $\boxed{a} \otimes \boxed{c} = \boxed{ac}$. Can you see why a prime is necessary if we are to have a group?

Now, for "More of the Same." Let's motivate this topic by asking, How many groups of order 2 are there?

Everyone know that there are infinitely many. Any set with 2 elements a and b forms a group by defining the binary operation by $aa = bb = a$ and $ab = ba = b$. However, an algebraist will quickly say that these are all the same and say that there is only 1 group of order 2. To be correct, the algebraist should add "up to isomorphism." It is this idea of sameness that we want to approach next.

Because multiplication of powers of a rotation in the plane is achieved simply by adding the exponents of the powers, the group C_n and the group $\mathbb{Z}_n$, the group of the integers under addition mod n, are "the same" as abstract groups. In this case, we say that the groups are "isomorphic," which literally means that they have the same shape. Formally, group G with binary operation $*$ is **isomorphic** to group H with binary operation $\star$ if there exist a one-to-one mapping π from G onto H such that

$$\pi(a * b) = \pi(a) \star \pi(b)$$

for all a and b in G. The mapping π, which is called an **isomorphism**, gives us the correspondence between the elements of one group with the elements of the other. (It is nice to think of this mapping as a Rosetta Stone.) The equation $\pi(a * b) = \pi(a) \star \pi(b)$ tells us that π translates the Cayley table for G to the Cayley table for H. In other words, the groups are the same except for notation. The mapping π given by $\pi(\rho^k) = k$ gives an isomorphism from C_n to $\mathbb{Z}_n$. An isomorphism from a group G onto G is called an **automorphism**. We may as well mention in passing that group G with binary operation $*$ is **homomorphic** to group H with binary operation $\star$ if there exist a mapping π from G into H such that $\pi(a * b) = \pi(a) \star \pi(b)$ for all a and b in G. To say that G and H are homomorphic is to say that they have "like" Cayley tables. So an isomorphism is a homomorphism that gives a one-to-one correspondence between the elements of one group and the elements of the other group.

How many groups of order 2 are there? A geometer has an answer different from the algebraist. For instance, there are 2 groups of order 2 among the isometries of the plane. The geometer says that groups are the same iff they are *conjugate*. (Details and definitions are given in The Back of the Book.) All the groups consisting of the identity ι and 1 rotation of 180° are conjugate to C_2 and are the same; all the groups consisting of ι and 1 reflection are conjugate to D_1, consisting of ι and the reflection in the X–axis, are the same. However, C_2 and D_1 are not conjugate and are deemed to be different, although isomorphic.

Homework.

1. Give 2 mappings such that each is an isomorphisms from D_3 to S_3.
2. Show that mapping π defined on C_{12} by $\pi(\rho^k) = \rho^{5k}$ is an automorphism on C_{12}.
3. Show that mapping π defined on C_{12} by $\pi(\rho^k) = \rho^{4k}$ is a homomorphism on C_{12}.

5
Actions

§37. The Definition

Group G with identity e **acts on** set X if for each element g in G there is an associated map π_g from X into X such that

 1. $\pi_e(x) = x$ for every x in X.

 2. $\pi_{hg}(x) = \pi_h(\pi_g(x))$ for every g and h in G and for every x in X.

The second axiom can be restated as $\pi_{hg} = \pi_h \pi_g$. Some immediate consequence of this short definition are explored in Exercise 1.

Exercise 1. With the notation in the definition above, show that

 1. π_g is a permutation on X, and $\pi_g^{-1} = \pi_{g^{-1}}$.

 2. $\pi_g(x) = y$ iff $x = \pi_{g^{-1}}(y)$.

 3. $\{\pi_g \mid g \in G\}$ is a group, which we denote by Π.

Note: Do not assume that $\pi_g \neq \pi_h$ just because $g \neq h$.

Those and, perhaps, only those who enjoy extremely concise definitions will appreciate observing that an action π of group G acting on set X could have been defined as a homomorphism from G into the group of all permutations on set X. In our notation, we would have $\pi(g) = \pi_g$. Surely, it is from the examples and problems below that we will gain an understanding of actions.

Most often, in our applications, group G acts on set X in a natural way, as follows. If G is a group of symmetries of a euclidean space E and X is a set of points or subsets of E, then we can define the **natural action** of G on X by the following:

$$\text{For each } g \text{ in } G, \text{ let } \pi_g(x) = g(x) \text{ for each } x \text{ in } X.$$

Thus, in this case, π_g and Π are just the restrictions of g and G, respectively, to the set X. It is this restriction Π that is a permutation group on X, as in the examples below. If X is a finite set, then π_g has a cycle type; if g is an isometry of euclidean space, then it doesn't make much sense to talk about the cycle type of g. We probably should verify that what we called the natural action above is truly an action of G on X according to our 2 axioms. This is easy since for each x in X we have (1) $\pi_e(x) = e(x) = x$ and (2) $\pi_{hg}(x) = hg(x) = h(g(x)) = \pi_h(\pi_g(x)) = \pi_h\pi_g(x)$.

We introduce the standard notation σ_l for the reflection in line l. Generally, we will take h to be the X-axis, take v the to be the Y-axis, take p to be the line with equation $Y = X$, and take m to be the line with equation $Y = -X$. Thus, the 8 elements of D_4 are then the 4 rotations ι, ρ, ρ^2, ρ^3 and the 4 reflections σ_h, σ_v, σ_p, σ_m. This notation is easy to use and easy to remember. (To tie this to the previous notation for D_4, with $\sigma = \sigma_h$, we have $\sigma\rho = \sigma_m$, $\sigma\rho^2 = \sigma_v$, and $\sigma\rho^3 = \sigma_p$.)

Let $G = D_4$, and let X be the set of the 4 quadrants $1, 2, 3, 4$ of the 2–by–2 checkerboard ⊞ centered at the origin. We take π to be the natural action of G on X. Then $\pi_\iota = (1)(2)(3)(4)$ and π_ι has cycle type z_1^4. Since $\pi_\rho = (1234)$ and $\pi_{\rho^{-1}} = (1432)$, then each of π_ρ and $\pi_{\rho^{-1}}$ has cycle type z_4^1. Since $\pi_{\rho^2} = (13)(24)$, then π_{ρ^2} has cycle type z_2^2. Each of π_{σ_h} and π_{σ_v} has cycle type z_2^2, since $\pi_{\sigma_h} = (14)(23)$ and $\pi_{\sigma_v} = (12)(34)$. Finally, each of π_{σ_p} and π_{σ_m} has cycle type $z_1^2 z_2^1$ since $\pi_{\sigma_p} = (1)(3)(24)$ and $\pi_{\sigma_m} = (13)(2)(4)$. The action is given by the π_g's. It is reasonable to ask, Why are all these cycle types given here? Alas, the answer is, That's not clear now, but be assured that whenever we compute elements of Π we will always want to have their cycle types.

Example. Let $G = D_3$, and let X be the set of edges of an equilateral triangle in standard position (the vertices are $(\cos 2\pi k/3, \sin 2\pi k/3)$), which are labeled counterclockwise from $(1,0)$: a, b, c. Then G acts on X in the natural way, as shown in Table 5.1. In passing, we note that for this example the 3 groups Π, S_3, and D_3 are isomorphic. Since $|D_n| = 2n$ and $|S_n| = n!$, then $D_n \cong S_n$ implies $n = 3$.

Homework.

1. Using the example above as a model, make a table similar to Table 5.1 for the group D_3 acting naturally on the set $\{A, B, C, D, E, F\}$ of vertices of a regular hexagon in standard position, where the vertices

Isometry g	π_g in cycle notation	Cycle type of π_g
ι	(a)	z_1^3
ρ	(abc)	z_3^1
ρ^2	(acb)	z_3^1
σ	(ac)	$z_1^1 z_2^1$
$\sigma\rho$	(ab)	$z_1^1 z_2^1$
$\sigma\rho^2$	(bc)	$z_1^1 z_2^1$

TABLE 5.1. D_3 Acting on Edges of an Equilateral Triangle.

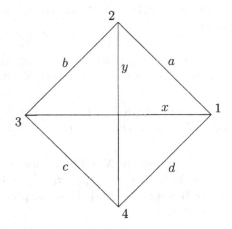

FIGURE 5.1. Notation for Square in Standard Position.

are listed counterclockwise around the hexagon with $A = (1, 0)$. (Yes, the hexagon does have (full) symmetry group D_6; however, here it is the group D_3 that is acting on the hexagon.)

2. Using the example above as a model, make similar tables for the group D_4 acting naturally on

   (a) the set $\{1, 2, 3, 4\}$ of the vertices of the square in Figure 5.1.

   (b) the set $\{a, b, c, d\}$ of the edges of the square in Figure 5.1.

   (c) the set $\{x, y\}$ of the diagonals of the square in Figure 5.1.

   (d) the set $\{a, b, c, d, x, y\}$ of diagonals and edges of the square in Figure 5.1.

3. Using the example above as a model, make a table similar to Table 5.1 for the group D_4 acting naturally on the set of the 16 colorings of the 2-by-2 checkerboard in Figure 5.2, where we suppose the checkerboards are centered at the origin. (Depending on how you think about the set X of colorings, the action here may or may not be technically

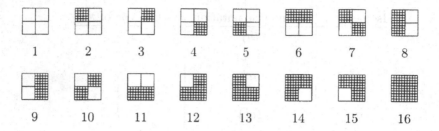

1   2   3   4   5   6   7   8

9   10   11   12   13   14   15   16

FIGURE 5.2. Colorings of the 2–by–2 Checkerboard.

natural by our definition; however, the "induced action" should be "obvious" in any case.)

We give 2 more examples of group actions that are important to algebraists. Both are of the form of group G acting on set X where $X = G$.

Example. Group G acts on set X where $X = G$ and $\pi_g(x) = gxg^{-1}$. (The right side of the equation makes sense since $x \in G$.) We call π_g an *inner automorphism* of G.

Example. Group G acts on set X where $X = G$, and for each g in G we have $\pi_g(x) = gx$ for each x in X. In this case, $\pi_g = \pi_h$ iff $g = h$, and so the permutation group Π, where $\Pi = \{\pi_g \mid g \in G\}$, is isomorphic to G. Thus, we have the following observation.

Cayley's Theorem. Every group is isomorphic to a permutation group. In particular, a finite group with n elements is isomorphic to a subgroup of S_n.

Although Cayley's Theorem is a beautiful result, we will not need the theorem in the sequel.

Exercise 2. In Table 5.2, fill in the Cayley table for the permutation group Π that is isomorphic to the group G, whose Cayley table is given in the table, and where the action is given by $\pi_g(h) = gh$, as in the example preceding Cayley's Theorem. (This is easy to do if you know what to do.) Also, argue that there are only 2 groups of order 4, up to isomorphism. The group G in Table 5.2 is called *Klein's Vierergruppe* or *Klein's four group*.

§38. Burnside's Lemma

We declare that 2-colorings of the 2–by–2 checkerboard are "the same" if one can be obtained from the other by the action of D_4 on the set of

G	e	a	b	c		Π	
e	e	a	b	c			
a	a	e	c	b			
b	b	c	e	a			
c	c	b	a	e			

TABLE 5.2. Cayley Tables for G and Π.

colorings in Figure 5.2. We partition these 16 colorings into 6 sets, where the 2-colorings are in the same set iff they are "the same," as defined above:

$$\{1\},\ \{2,3,4,5\},\ \{6,8,9,11\},\ \{7,10\},\ \{12,13,14,15\},\ \{16\}.$$

We conclude that there are exactly 6 ways to 2-color the vertices of a square that is "free in space." We found the 6 sets above easily by hunting them out from our homework. These 6 equivalence classes determined by the partition and displayed above are more obvious if we replace the numerals by pictures to get

$$\{\boxplus\},\ \{\blacksquare,\blacksquare,\blacksquare,\blacksquare\},\ \{\blacksquare,\blacksquare,\blacksquare,\blacksquare\},\ \{\blacksquare,\blacksquare\},\ \{\blacksquare,\blacksquare,\blacksquare,\blacksquare\},\ \{\blacksquare\}.$$

However, for the 2-colorings of a regular 8-by-8 checkerboard, we would have to look at 2^{64} colorings. That's a bit much. We need to develop some theory to handle such problems, including the special problem, How many ways are there to 2-color the 64 squares in an 8–by–8 checkerboard? Of course, before an answer can be given, we must agree on when colorings are "the same." Here, there are at least 2 reasonable answers. We consider it reasonable to have D_4 acting on the colorings, while at other times we might consider only C_4.

Lemma 1. Suppose group G acts on set X. If relation $\sim$ is defined on X by $x \sim y$ iff there exists g in G such that $\pi_g(x) = y$, then this relation is an equivalence relation.

Proof. We suppose G has identity e. We have

$$\sim\ = \{(x,y) \mid x,y \in X \text{ and } \exists g \in G \ni \pi_g(x) = y\}.$$

We verify the 3 properties defining an equivalence relation. (R): $x \sim x$ because $\pi_e(x) = x$. (S): Suppose $x \sim y$. Then there is a g in G such that $\pi_g(x) = y$. So, $x = \pi_g^{-1}\pi_g(x) = \pi_{g^{-1}}\pi_g(x) = \pi_{g^{-1}}(y)$, proving that $y \sim x$. (T): Suppose that $x \sim y$ and $y \sim z$. Then there are g and h in G such that $\pi_g(x) = y$ and $\pi_h(y) = z$. Then, since $\pi_{hg}(x) = \pi_h\pi_g(x) = \pi_h(y) = z$, we have $x \sim z$. ∎

The equivalence classes of the equivalence relation in the lemma are called the **orbits** of the group action. For x in X, denote the orbit of x by

O_x. Hence, $O_x = O_y$ iff $x \sim y$ and

$$O_x = \{y \mid x \sim y\}$$
$$= \{y \mid y = \pi_g(x) \text{ for some g in } G\}$$
$$= \{\pi_g(x) \mid g \in G\}.$$

Note that "O_x" is a more suggestive notation than our former, generic notation "$\boxed{x}$" from the previous chapter. For each x in X, let

$$S_x = \{g \in G \mid \pi_g(x) = x\}.$$

We call S_x the **stabilizer** of x in X. It is very important to note that O_x is a subset of X, while S_x is a subset of G.

Lemma 2. If group G acts on set X, then the stabilizer of x in X is a subgroup of G

Proof. We suppose G has identity e and x is in X. We want to show that S_x is a subgroup of G. We verify the axioms for a group. If g and h are in S_x, then $\pi_{hg}(x) = \pi_h \pi_g(x) = \pi_h(\pi_g(x)) = \pi_h(x) = x$, proving that hg is in S_x. So, S_x is closed. Also, S_x is associative since G is associative. Further, e is in S_x because $\pi_e(x) = x$. Finally, since $\pi_g(x) = x$ iff $x = \pi_{g^{-1}}(x)$, then $\pi_g \in S_x$ iff $\pi_{g^{-1}} \in S_x$. Hence, S_x has the inverse property. ∎

Exercise. For each of the 4 actions from Exercise 2 on page 87, form the table with columns headed by elements in X and 3 rows headed by x, $|O_x|$, $|S_x|$. Also, form a similar table for the action from Exercise 3 on page 87.

Lemma 3. With the notation above, for a fixed x in X there is a one-to-one correspondence between the elements of O_x and the set of all cosets of S_x in G.

Proof. If G is finite, we want to show that there is a one-to-one correspondence between the elements

$$\pi_{g_1}(x), \ \pi_{g_2}(x), \ \pi_{g_3}(x), \ \dots, \ \pi_{g_k}(x)$$

of O_x and the set of all cosets

$$g_1 S_x, \ g_2 S_x, \ g_3 S_x, \ \dots, \ g_k S_x$$

of S_x in G. Whether G is finite or not, let f be the mapping from O_x to the set of cosets of S_x that is defined by $f(\pi_g(x)) = g S_x$. Mapping f is clearly onto, since the coset $g S_x$ is the image of $\pi_g(x)$. To show that f is one-to-one, suppose $g S_x = h S_x$. Then $h^{-1} g S_x = S_x$ and so $h^{-1} g$ is in S_x. Hence $\pi_{h^{-1} g}(x) = x$ and $\pi_g(x) = \pi_h(x)$, as desired. ∎

Lemma 4. With the notation above, if x is in X then

$$|O_x| \cdot |S_x| = |G|.$$

Proof. $|G| = |S_x| \cdot |G : S_x| = |S_x| \cdot |O_x|$ by the preceding lemma. ∎

Lemma 5. With the notation above, if t is in X, then $O_x = O_t$ for every x in O_t.

Proof. $O_x = O_t$ because equivalence classes having an element in common are identical. ∎

Lemma 6. With the notation above, if t is in X then

$$\sum_{x \in O_t} |S_x| = |G|.$$

Proof. If $x \in O_t$, then $O_x = O_t$. So, by the preceding 2 lemmas, we have

$$\sum_{x \in O_t} |S_x| = \sum_{x \in O_t} \frac{|G|}{|O_x|} = \sum_{x \in O_t} \frac{|G|}{|O_t|} = |O_t| \cdot \frac{|G|}{|O_t|} = |G|. \blacksquare$$

Lemma 7. With the notation above, if ω is the number of orbits of group G acting on set X, then

$$\omega = \frac{1}{|G|} \sum_{x \in X} |S_x|.$$

Proof. By the preceding lemma, we have

$$\sum_{x \in X} |S_x| = \omega \cdot |G|,$$

since each x is in exactly 1 orbit, which is the desired formula. ∎

Applying this lemma to our 2-coloring of a 2–by–2 checkerboard, we have $\frac{1}{|G|} \sum_{x \in X} |S_x| = \frac{48}{8} = 6$. Although the answer is correct of course, this formula is not good enough to be useful because, in general, X is too large. For example, suppose X is the set of the 2^{64} possible 2-colorings of an 8–by–8 checkerboard in a fixed position. Since G is usually much smaller than X, it would be nice to have a count based on G rather than X.

We introduce the **fix** of each element g in G by

$$F_g = \{x \in X \mid \pi_g(x) = x\}.$$

Note that F_g is a subset of X, while S_x is a subgroup of G.

Lemma 8. With the notation above,

$$\sum_{g \in G} |F_g| = \sum_{x \in X} |S_x|.$$

Proof. Here, we use the very valuable technique of counting something in 2 ways and setting the calculations equal to each other. We are counting the number of elements in the set of all ordered pairs (x, g) with $x \in X$ and $g \in G$ such that $\pi_g(x) = x$. First, for each x in X, we have $|S_x|$ ordered pairs. So, all together, we must have $\sum_{x \in X} |S_x|$ pairs. On the other hand, for each g in G we have $|F_g|$ pairs. So, all together, we must have $\sum_{g \in G} |F_g|$ ordered pairs. Setting the 2 calculations equal to each other gives the desired equation. ∎

Our important theorem now follows directly from the last 2 lemmas. Yes, Burnside's Lemma was anticipated by other mathematicians. However, since most mathematical names applied to mathematical results are questionable, we will stick with the traditional name.

Theorem *Burnside's Lemma*. If group G act on set X, then the number of orbits is

$$\frac{1}{|G|} \sum_{g \in G} |F_g|$$

where $F_g = \{x \in X \mid \pi_g(x) = x\}$.

Let's apply Burnside's Lemma to Figure 5.2 and our 2-coloring of the 2–by–2 checkerboard. We compute the following array, where the top row lists all the elements of D_4 and the bottom row is determined by placing under isometry g in the top row the number of the 16 colorings in Figure 5.2 that are fixed (left unchanged) by permutation π_g. That is, underneath g in the top row we enter $|F_g|$ in the bottom row. For example, since the permutation π_{ρ^2} fixes exactly the 4 colorings ⊞, ◧, ◨, ■, which are numbered $1, 7, 10, 16$ in Figure 5.2, then under ρ^2 in the array we have the entry 4. The complete array is

$$\begin{pmatrix} \iota & \rho & \rho^2 & \rho^3 & \sigma_h & \sigma_v & \sigma_p & \sigma_m \\ 16 & 2 & 4 & 2 & 4 & 4 & 8 & 8 \end{pmatrix}.$$

By Burnside's Lemma, it follows that

$$\omega = \frac{16 + 2 + 4 + 2 + 4 + 4 + 8 + 8}{8} = 6,$$

as expected. If we consider only C_4 acting on the set of colorings, we will get the same result, $\omega = 6$. It would not be wise to generalize, however.

The first problem in the homework below asks about 2-coloring the 8–by–8 checkerboard. See Figure 5.3. We can hardly apply the technique that we just used above for the 2–by–2 checkerboard to this problem. Here there are 2^{64} possible colorings to consider. We will take another pass at the 2–by–2 checkerboard problem, but this time we will suppose that we do not have Figure 5.2 available. For each g in G, we should be able to calculate the value of $|F_g|$ by considering only the figure ⊞. That is, we

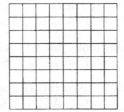

FIGURE 5.3. The 8–by–8 Checkerboard.

ask ourself for each g in G, How many ways are there to 2-color the 2–by–2 checkerboard ⊞, such that the resulting pattern has g as a symmetry? For example, suppose $g = \rho^2$. In this case, the 2 squares in the first and third quadrants must be the same color, since these 2 squares are interchanged by ρ^2. Likewise, the 2 squares in the second and fourth quadrants must also be the same color. So, we have $2 \cdot 2$ ways to color the squares so that the coloring is fixed by ρ^2. Another possible way to count these is to see that the top 2 squares can be colored in $2 \cdot 2$ ways and then the coloring of the bottom 2 squares is determined by ρ^2. For another example, suppose $g = \sigma_p$. In this case, since σ_p interchanges the 2 squares in the second and fourth quadrants, these squares must be the same color. However the other 2 squares can each be colored in 2 ways since each is fixed by σ_p. Hence, here we have 2^3 colorings that are fixed by σ_p. For another view, each of the 3 squares marked with a dot in the figure ⊞ can be colored in 2 ways and the color of the others (only 1 square here) is determined by the symmetry. Of all possible 2^4 colorings of ⊞, permutation π_{σ_p} fixes exactly these 2^3 colorings and permutes the other colorings among themselves. Without any reference to Figure 5.2, this time we get the array

$$\begin{pmatrix} \iota & \rho & \rho^2 & \rho^3 & \sigma_h & \sigma_v & \sigma_p & \sigma_m \\ 2^4 & 2^1 & 2^2 & 2^1 & 2^2 & 2^2 & 2^3 & 2^3 \end{pmatrix}.$$

Of course, $\omega = 6$, as before. This technique will be very helpful in homework problems below. We will generally use ω to denote the number of orbits of G acting on X.

Our 2 computations of the result $\omega = 6$ are based on different views. In the first, we found F_g and then calculated $|F_g|$. The shift in the second calculation allows us to count the number of elements in F_g without ever listing them. For example, we calculated that there are 2^3 colorings that are fixed by π_{σ_p} without explicitly determining these colorings. In the second view, we compute $|F_g|$ without computing F_g. That saves a lot of work.

Homework.

1. How many ways can we 2-color the 64 squares in the 8–by–8 checkerboard, with C_4 acting on the colorings? See Figure 5.3.

2. How many ways can we 2-color the 64 squares in the 8–by 8 checkerboard, with D_4 acting on the colorings? See Figure 5.3.

3. How many ways can we 3-color the 64 squares in the 8–by–8 checkerboard, with D_4 acting on the colorings? See Figure 5.3.

4. How many ways can we m-color the 64 squares in the 8–by–8 checkerboard, with D_4 acting on the colorings? See Figure 5.3.

5. How many ways can we m-color the 49 squares in the 7–by–7 checkerboard, with D_4 acting on the colorings? See Figure 5.4.

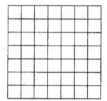

FIGURE 5.4. The 7–by–7 Checkerboard.

6. How many necklaces of 6 stones (use D_6) can be made from rubies, diamonds, and emeralds? (For a less romantic version of the problem, consider counting the 6-bead necklaces that can be made with spherical beads, each having 1 of 3 possible colors. At least in mathematics, all necklaces can be turned over and have no clasp. In other words, we can consider a **necklace problem** as coloring the vertices of a regular n-gon and use D_n as the group of the action.)

7. How many ways, under D_8, can we paint the 8 spokes of a wheel with 4 available colors if each spoke is all 1 color?

§39. Applications of Burnside's Lemma

Burnside's Lemma is a very powerful theorem. This section consists mainly of problems, most of which can not be done without the investment of some time.

We have been calculating the cycle types for various permutations in Π without having any particular use for the result. We now introduce their mean for a given action. It is still not evident why we should be interested in this average; however, we are.

The **cycle index**, traditionally denoted as P_G, for group G acting on set X is the average (mean) of the $|G|$ cycle types of π_g for g in G. So, if group G acts on set X, to get the cycle index we sum the cycle types, 1 for each element in G, and then divide the sum by $|G|$. For example, for D_3 acting on the edges of an equilateral triangle, we compute the cycle index to get

$$P_G = \frac{z_1^3 + z_3^1 + z_3^1 + z_1^1 z_2^1 + z_1^1 z_1^1 + z_1^1 z_2^1}{6}$$

directly from Table 5.1 on page 87.

Example. We turn to a completely nontrivial problem that illustrates the power of our theory. How many ways are there to m-color the faces of a regular dodecahedron?

We suppose π is the natural action of the icosahedral group G of all 60 rotation symmetries of the dodecahedron acting on the set X of the 12 faces. Recall that a regular dodecahedron has 12 faces, 30 edges, and 20 vertices. As will now be our customary notation, X^* then denotes the set of m-colorings of X and π^* is the natural action of G on X^*. We are looking for ω^*.

Study Table 5.3 until you completely understand each and every entry. Do not fool yourself; this will take some time. Note that the footnotes at the bottom of the table makes sense only for the action π of G acting on the set of *faces*, numbered 1–12. Our initial task is to compute $|F_g^*|$, the number of elements in the fix of g in the action π^* of the group G acting on the set of *colorings* of the faces, for each g in G. Then we compute the cycle type for each g in G for G acting on the *faces*. (Here, $|X| = 12$ and $|X^*| = m^{12}$. Note that for G acting on the faces we have $\omega = 1$, since any face can be brought to any other by a rotation. So, ω is not very interesting. We are, however, interested in ω^*, the number of equivalence classes for G acting on the set of colorings. It is ω^* that is the number of different colorings of the faces of the dodecahedron.)

After studying Table 5.3, we see that

$$\omega^* = \frac{1 \cdot m^{12} + 24 \cdot m^4 + 15 \cdot m^6 + 20 \cdot m^4}{60}$$

by Burnside's Lemma. The cycle index for the action on the faces is given by

$$P_G(z_1, z_2, z_3, z_4, z_5) = \frac{1 \cdot z_1^{12} + 24 \cdot z_1^2 z_5^2 + 15 \cdot z_2^6 + 20 \cdot z_3^4}{60}.$$

It has probably not escaped your attention that

$$\omega^* = P_G(m, m, m, m, m).$$

Since the z_i's are indeterminates, it makes absolutely no sense to say, for example, "let $z_2 = m$," but it does make sense to say "replace z_2 in P_G by m." The difference is subtle but necessary. Thus, we understand what "$P_G(m, m, m, m, m)$" means.

# Symmetry g in G	# $|F_g^*|$ for G acting on the colorings of the faces	# Cycle type of π_g for G acting on the faces
1 identity*	1 m^{12}	1 z_1^{12}
$6 \cdot 4$ rotations of order 5 (about the join of centers of opposite faces)**	24 m^4	24 $z_1^2 z_5^2$
$15 \cdot 1$ rotations of $180°$ (about join of centers of opposite edges)***	15 m^6	15 z_2^6
$10 \cdot 2$ rotations of $120°$ (about join of centers of opposite vertices)****	20 m^4	20 z_3^4

*	So $\pi_\iota = (1)(2)(3)(4)(5)(6)(7)(8)(9)(10)(11)(12)$.
**	E.g.: $\pi_\alpha = (1)(2,3,4,5,6)(7,8,9,10,11)(12)$.
***	E.g.: $\pi_\beta = (1,2)(3,4)(5,6)(7,8)(9,10)(11,12)$.
****	E.g.: $\pi_\gamma = (1,2,3)(4,7,5)(6,8,9)(10,11,12)$.

TABLE 5.3. Coloring the Faces of a Dodecahedron.

How do we explain this marvelous coincidence? Let's consider the entry m^4 for α in Table 5.3. We take face 1 as the top face and face 12 as the bottom face of the dodecahedron. Under π_α, the top face goes to itself and the bottom face goes to itself. So each of these 2 faces can be any of the m colors. That is, we can color these 2 faces in m^2 ways. The 5 faces adjacent to the top face cycle among themselves under π_α. Hence, all these faces must be colored with the same color if the coloring of the dodecahedron is left unchanged; we have m choices for this color. Likewise, the 5 faces adjacent to the bottom side must be of the same color, and we have m choices for this color. All together, we have m^4 choices for each of the 24 rotations of order 5 in G. Comparing the columns of Table 5.3, we see that the middle column is obtained by replacing each z_i in the right column by m. In fact, 2 faces must be of the same color if they are in the same cycle of π_α. Since π_α is the product of 4 disjoint cycles, then $|F_\alpha^*| = m^4$. In general the cycle type allows us to find $|F_g^*|$, without having to compute F_g^*. Apparently, the cycle types are a very useful tool to have on a picnic.

You might object and say that we do not need all these indeterminates because we use only the exponents and do nothing with the subscripts. The exponents count the number of cycles and that is all we really need to know. That is true, for now, but rest assured, we will have a use for the subscripts as well as the exponents. In any case, if we know the cycle structure of π_g, then it easy to calculate $|F_g^*|$, the number of elements in the fix of g, when the action is on the colorings. In summary, our very important equation

$$\omega^* = P_G(m, m, m, m, m)$$

is an immediate consequence of Burnside's Lemma and the following.

Observation. With all our previous notation, if permutation π_g is a product of k disjoint cycles, then $|F_g^*| = m^k$.

Homework 1.

Recall that the octahedral group is the rotation group of the cube. In the 3 figures in Figure 5.5, typical rotations are indicated by their axes. In particular, α is a rotation of order 4 about a join of 2 centers of opposite faces, β is a rotation of order 2 about the join of 2 midpoints of opposite edges, and γ is a rotation of order 3 about the join of 2 opposite vertices. As for Figure 5.5, it helps to know that the sum of the numbers on opposite faces of a die is always 7. So π_β interchanges faces 1 and 4. These figures are given so that a class will all be using the same notation. Of course, the cycle index will be the same whatever notation is chosen.

1. Under the octahedral group, how many ways are there to m-color the faces of the cube? (Use the notation in Figure 5.5.)

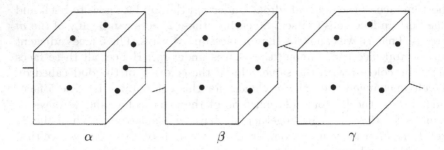

FIGURE 5.5. Symmetries of a Cube Acting on Faces.

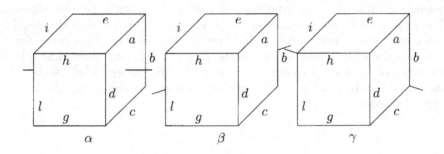

FIGURE 5.6. Symmetries of a Cube Acting on Edges.

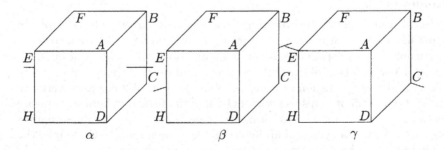

FIGURE 5.7. Symmetries of a Cube Acting on Vertices.

2. Under the octahedral group, how many ways are there to m-color the edges of the cube? (Use the notation in Figure 5.6.)

3. Under the octahedral group, how many ways are there to m-color the vertices of the cube? (Use the notation in Figure 5.7.)

Homework 2.

1. How many different 2–by–2–by–2 cubes can be constructed from 8 of an unlimited number of unit cubes of 5 different colors?

2. How many ways are there to arrange 2 M's, 4 A's, 5 T's and 6 H's under the condition that any arrangement and its reverse order are considered the same?

3. How many ways can the faces of a cube be colored if each face is colored 1 of 6 colors and each color is used?

4. How many ways can each of the 2 figures in Figure 5.8 be m-colored? (Obviously, each disk is to be colored with exactly 1 of the m colors. Not obviously, consider colorings to be the same if one can be obtained from the other by the action on the colorings induced by D_3.)

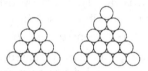

FIGURE 5.8. Billiard Balls.

5. How many ways, under D_4, are there to 3-color the 60 edges in Figure 5.9.

FIGURE 5.9. 5–by–5 Checkerboard.

6. How many ways are there to 5-color the 18 edges in Figure 5.10 under the rotation group of the figure?

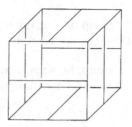

FIGURE 5.10. Marked Cube.

§40. Pólya's Pattern Inventory

Let's begin by formalizing something that we know implicitly from the very beginning of our study of counting. Suppose that $i, j, \ldots, k$, and n are nonnegative integers and that $i + j + \cdots + k = n$. From the expression

$$\overbrace{(A + B + \cdots + Z)(A + B + \cdots + Z) \cdots (A + B + \cdots + Z)}^{n \text{ factors}},$$

we see that in the expansion of $(A + B + \cdots + Z)^n$ without collecting terms there are just as many terms $A^i B^j \cdots Z^k$ as there are arrangements of i A's, j B's, $\ldots$, and k Z's, since each term in the expansion is formed by selecting 1 term from each of the n factors. We know the number of arrangements, as a Mississippi problem, is $n!/(i!j! \cdots k!)$. Another approach has us first selecting i of the factors to contribute an A in $\binom{n}{i}$ ways. Then, we select j of the remaining $n - i$ factors to contribute a B in $\binom{n-i}{j}$ ways. $\ldots$ Finally, we select k of the remaining k factors to contribute a Z in $\binom{k}{k}$ ways. The product of all these binomial coefficients gives the same $n!/(i!j! \cdots k!)$. In either case, we have the following.

Observation. Suppose $i, j, \ldots, k$ and n are nonnegative integers. Then the coefficient of $A^i B^j \cdots Z^k$ in $(A + B + \cdots + Z)^n$ is

$$\frac{n!}{i!j! \cdots k!}$$

if $i + j + \cdots + k = n$ and 0 otherwise.

The observation above or the equivalent equation

$$(A + B + \cdots + Z)^n = \sum_{i+j+\cdots+k=n} \frac{n!}{i!j! \cdots k!} A^i B^j \cdots Z^k$$

is called the **Multinomial Theorem**. The Binomial Theorem is a special case of the Multinomial Theorem.

Let's return to our example of coloring the faces of a dodecahedron and Table 5.3 on page 96, but now suppose that $m = 3$ and that the 3 colors are, what else?: red, white, and blue. No problem so far. However, suppose we are interested in finding the number of colorings (under the icosahedral group) of the faces of the dodecahedron with the restriction that there are 2 red faces, 4 white faces, and 6 blue faces. This restriction is a big problem, but nothing we cannot handle. We add a new column at the right of Table 5.3, with the heading $|F'_g|$ for g in group G acting on the set X' of our *restricted* colorings, that is, the set of those colorings with the restriction that there are 2 red faces, 4 white faces, and 6 blue faces. We will stick with our basic notation but add a prime to distinguish this action on the restricted colorings. So G is the icosahedral group, π is the natural action of G on the set X of the faces of the docecahedron, and π' is the natural action of G on the set X' of colorings of the dodecahedron that have 2 red faces, 4 white faces, and 6 blue faces. We will look at the 4 representative rotations ι, α, β, γ, one at a time, and recall a little of what we know about generating functions. (Surprise!) As we go along, compare the cycle type column in Table 5.3 with our new column.

Given the rotation ι where $\pi_\iota = (1)(2)(3)(4)(5)(6)(7)(8)(9)(10)(11)(12)$, how many of the restricted colorings are fixed by π'_ι? The generating function for coloring any 1 of the 12 faces is $r^1 + w^1 + b^1$ and so the generating function for coloring all 12 of the faces is $(r^1 + w^1 + b^1)^{12}$, where you can guess the reason for the choice of indeterminates r, w, b. Therefore, we are looking for the coefficient of $r^2 w^4 b^6$ in $(r+w+b)^{12}$, which is $12!/(2!4!6!)$, or 13860, and which we knew as the answer in the first place without invoking generation functions. This is, after all, only a simple Mississippi Problem; is it not? (How many 12-letter words can be made from 2 r's, 4 w's, and 6 b's?) However, generating functions will be useful in the less obvious cases considered next.

Given the rotation α where $\pi_\alpha = (1)(2,3,4,5,6)(7,8,9,10,11)(12)$, how many of the restricted colorings are fixed by π'_α? The generating function for coloring face 1 is $r^1 + w^1 + b^1$. The cycle $(2,3,4,5,6)$ in π_α tells us that face 3 must have the same color as face 2, that face 4 must have the same color as face 3, that face 5 must have the same color as face 4, that face 6 must have the same color as face 2, and that face 2 must have the same color as face 6. Since these 5 faces must all have the same color, the generating function for coloring the set of these 5 faces is $r^5 + w^5 + b^5$. Likewise, the colorings of the individual faces in the cycle $(7,8,9,10,11)$ are not independent. We must again consider the set of all 5 faces at once, and again we have the generating function $r^5 + w^5 + b^5$. The coloring of face 12 is independent of all previous colorings, and the generating function for coloring this face is simply $r^1 + w^1 + b^1$. Therefore, the generating function for coloring all the faces here is $(r + w + b)^2 (r^5 + w^5 + b^5)^2$. Thus, our entry, in our new column, that is to the right of $z_1^2 z_5^2$ in Table 5.3 is the coefficient of $r^2 w^4 b^6$ in $(r + w + b)^2 (r^5 + w^5 + b^5)^2$. This coefficient is 0.

Given the rotation β where $\pi_\beta = (1,2)(3,4)(5,6)(7,8)(9,10)(11,12)$, how many of the restricted colorings are fixed by π'_β? In order to get independent generating functions, we must consider the faces in the pairs that are determined by the cycles in π_β. For each of the 6 transpositions, the generating function is $r^2 + w^2 + b^2$. In this case, the generating function is $(r^2 + w^2 + b^2)^6$ and we are looking for the coefficient of $r^2 w^4 b^6$ in $(r^2 + w^2 + b^2)^6$, which we may think of as the coefficient of $[r^2]^1 [w^2]^2 [b^2]^3$ in $([r^2] + [w^2] + [b^2])^6$ and which is $6!/(1!2!3!)$, or 60.

Given the rotation γ where $\pi_\gamma = (1,2,3)(4,7,5)(6,8,9)(10,11,12)$, how many of the restricted colorings are fixed by π'_γ? For each of the 4 3-cycles in π_γ, the generating function is $r^3 + w^3 + b^3$. In this case, the generating function is $(r^3 + w^3 + b^3)^4$, and we are looking for the coefficient of $r^2 w^4 b^6$ in $(r^3 + w^3 + b^3)^4$. This coefficient is 0.

Now, the answer to our specific problem with 2 red faces, 4 white faces, and 6 blue faces is, by Burnside's Lemma,

$$\frac{1(13860) + 24(0) + 15(60) + 20(0)}{60},$$

which is 246. However, much more important is the technique that we have observed and which we can apply to other problems. Comparing the cycle type for each of the representative permutations and the corresponding generating functions, we see that all the information that we needed is in the cycle type of the action of the group on the faces. In each case, the generating function is obtained by replacing each z_i in the cycle type by $r^i + w^i + b^i$. (At last, we begin to fully understand the importance of the cycle types.) Thus, by Burnside's Lemma, we want the coefficient of $r^2 w^4 b^6$ in the right-hand side of

$$P_G(z_1, z_2, z_3, z_4, z_5) = \frac{1 \cdot z_1^{12} + 24 \cdot z_1^2 z_5^2 + 15 \cdot z_2^6 + 20 \cdot z_3^4}{60}$$

after each z_i has been replaced by $r^i + w^i + b^i$. The answer to our specific problem is then the coefficient of $r^2 w^4 b^6$ in the expression

$$P_G(r + w + b, r^2 + w^2 + b^2, r^3 + w^3 + b^3, r^4 + w^4 + b^4, r^5 + w^5 + b^5),$$

or

$$\frac{[1](r + w + b)^{12} + [24](r + w + b)^2 (r^5 + w^5 + b^5)^2 + [15](r^2 w^2 b^2)^6 + [20](r^3 + w^3 + b^3)^4}{60}.$$

In general, for i red faces, j white faces, and k blue faces, where $i + j + k = 12$, we want the coefficient of $r^i w^j b^k$ in the expression above. Conversely, we have a generating function for the number of colorings with i red faces, j white faces, and k blue faces because the coefficient of $r^i w^j b^k$ in the expression above is precisely the number of different colorings (under the icosahedral group) having i red faces, j white faces, and k blue faces. For this reason, this generating function is called a **pattern inventory** or **Pólya's pattern inventory**.

Suppose the 2 colors b (for *black*) and w (for *white*) are available for coloring the vertices A, B, C, D of a square. Let $X = \{A, B, C, D\}$. "To color X" means that we assign to each element in X an element of $\{b, w\}$. Conversely, any mapping p from X into $\{b, w\}$ is a "coloring" of X. For example, suppose $p(A) = b$ and $p(B) = p(C) = p(D) = w$. This mapping p, which we can write as $\left(\begin{smallmatrix} A & B & C & D \\ b & w & w & w \end{smallmatrix}\right)$, essentially says "paint A black and paint each of B, C, D white." Our next definition generalizes and formalizes this simple idea.

An m-**coloring** of set X is a map from X into a nonempty set C where $C = \{c_1, c_2, \ldots, c_m\}$ and the elements of C are called **colors**. With C fixed, we generally let X^* denote the the set of all colorings of X with C. Rather than assign each color an indeterminate, it simplifies the notation if we suppose that the colors are themselves indeterminates, and we will do so.

With this notation and our usual notation for action π of group G on set X, we say that colorings p and q in X^* are **equivalent under** G (acting on X) and write $p \overset{*}{\sim} q$ iff there exists a g in G such that $p = q\pi_g$. (Pause and think about this. It *is* what we want. Note that $q\pi_g$ is defined but that $\pi_g q$ is not.) In order to use our theory, we need an action π^* of G on X^* such that $\overset{\pi^*}{\sim} = \overset{*}{\sim}$. Then $\overset{*}{\sim}$ is an equivalence relation and the number ω^* of equivalence classes is the number of ways to color X with C under G. We claim the following does the job. For each g in G, define an associated map π_g^* from X^* into X^* by

$$\pi_g^*(p) = p\pi_g^{-1}$$

for every p in X^*. As we will see, the bothersome -1 in our definition of π_g^* is necessary to get things in the right order, since the inverse of a product in a group is the product of the inverses in reverse order.

We interrupt the development of our theory for an illustration. We return to 2-coloring the 2–by–2 checkerboard ⊞. Here, group D_4 is acting on the set X of the 4 squares of the checkerboard, named $1, 2, 3, 4$ after the quadrants. So, $X = \{1, 2, 3, 4\}$, and the set of colors is given by $C = \{b, w\}$. We will use boldface numerals for the colorings suggested by Table 5.2. So, for example, the coloring **6** of the 2–by–2 checkerboard is the mapping $\left(\begin{smallmatrix} 1 & 2 & 3 & 4 \\ b & b & w & w \end{smallmatrix}\right)$, which corresponds to the picture ▥. Coloring **6** should be "the same" as coloring **8**, which is the mapping $\left(\begin{smallmatrix} 1 & 2 & 3 & 4 \\ w & b & b & w \end{smallmatrix}\right)$ that corresponds to the picture ▧. For the rotation ρ of 90°, we have π_ρ is (1234), or $\left(\begin{smallmatrix} 1 & 2 & 3 & 4 \\ 2 & 3 & 4 & 1 \end{smallmatrix}\right)$. We do indeed have **6** $\overset{*}{\sim}$ **8** since

$$\pi_\rho^*(\mathbf{6}) = \mathbf{6}\pi_\rho^{-1} = \begin{pmatrix} 1 & 2 & 3 & 4 \\ b & b & w & w \end{pmatrix} \begin{pmatrix} 1 & 2 & 3 & 4 \\ 4 & 1 & 2 & 3 \end{pmatrix} = \begin{pmatrix} 1 & 2 & 3 & 4 \\ w & b & b & w \end{pmatrix} = \mathbf{8}.$$

(Perhaps we should have previously called a picture such as ▉ a colored checkerboard rather than a coloring of the checkerboard. We doubt that this has lead to any confusion though. Supposedly, now that a *coloring* has been formally defined as a mapping, we should use the term only with its technical meaning.)

We now prove that π^*, which maps g to π^*_g, is an action of G on X^* by checking the 2 axioms at the beginning of the chapter. Certainly, with e the identity of G, we have $\pi^*_e(p) = p\pi_e^{-1} = p$ for every p in X^*. Also, for every g and h in G and for every p in X^*, we have $\pi^*_{hg}(p) = p\pi_{hg}^{-1} = p(\pi_h\pi_g)^{-1} = p\pi_g^{-1}\pi_h^{-1} = \pi^*_h(p\pi_g^{-1}) = \pi^*_h(\pi^*_g(p)), = \pi^*_h\pi^*_g(p)$, as desired.

We have the equivalence relation $\overset{\pi^*}{\sim}$ for the action π^* of G on X^*. Since, for any colorings p and q of X,

$$p \overset{\pi^*}{\sim} q \ \text{ iff } \ \pi^*_g(p) = q \text{ for some } g \text{ in } G$$

$$\text{iff } \ p\pi_g^{-1} = q \text{ for some } g \text{ in } G$$

$$\text{iff } \ p = q\pi_g \text{ for some } g \text{ in } G$$

$$\text{iff } \ p \overset{*}{\sim} q,$$

then $\overset{\pi^*}{\sim} = \overset{*}{\sim}$, as desired. The number of different (under G) colorings is the number of equivalence classes, which is, by Burnside's Lemma,

$$\frac{1}{|G|} \sum_{g\in G} |F^*_g| \text{ where } F^*_g = \{p \in X^* \mid \pi^*_g(p) = p\}$$

$$= \{p \in X^* \mid p = p\pi_g\} \text{ for } g \text{ in } G.$$

The result below then follows from the very important observation that $p = p\pi_g$ iff all the elements of X in each one of the cycles of π_g get assigned the same color by p.

Pólya's Pattern Theorem. If action π of group G on set X has cycle index $P_G(z_1, z_2, \ldots, z_n)$, if X^* is the set of m-colorings of X with set C of colors where $C = \{c_1, c_2, \ldots, c_m\}$, and if the action π^* of G on X^* is defined by $\pi^*_g(p) = p\pi_g^{-1}$ for p in X^* and g in G, then the number of different (under G) m-colorings of X is

$$P_G(m, m, \ldots, m)$$

and the pattern inventory of the m-colorings of X under G with colors C is

$$P_G(c_1 + c_2 + \cdots + c_m, c_1^2 + c_2^2 + \cdots + c_m^2, c_1^3 + c_2^3 + \cdots + c_m^3, \ldots, c_1^n + c_2^n + \cdots + c_m^n).$$

It can be mentioned in passing that Pólya was neither the first nor the last to develop this theory.

Example. How many ways are there to 5-color the edges of a regular dodecahedron? How many ways are there to 5-color the edges of a regular dodecahedron if exactly 5 edges are green?

We first compute the cycle index

$$\frac{z_1^{30} + 15z_1^2 z_2^{14} + 20z_3^{10} + 24z_5^6}{60}$$

for the icosahedral group acting on the set of edges of the regular dodecahedron.

Replacing each z_i in the cycle index by 5 then gives us the humongous number 15,522,042,948,408,209,375 as the answer to the first question.

For the second question we need the pattern inventory, which is

$$\big[(r + w + b + v + g)^{30}$$
$$+ 15(r + w + b + v + g)^2(r^2 + w^2 + b^2 + v^2 + g^2)^{14}$$
$$+ 20(r^3 + w^3 + b^3 + v^3 + g^3)^{10} + 24\ (r^5 + w^5 + b^5 + v^5 + g^5)^6\big] /60,$$

if the colors are r, w, b, v, g. We desire the sum of all the coefficients in the expansion of the pattern inventory that are of the form g^5 times a product of powers of the other 4 colors. Let's look at one term at a time in the numerator of the pattern inventory. In the product of 30 of the factors $(r + w + b + v + g)$, we must pick 5 of the factors to contribute a g in $\binom{30}{5}$ ways; and from each of the remaining 25 factors, we must select 1 of r, w, b, v in 4 ways. In the product

$$(r + w + b + v + g)^2(r^2 + w^2 + b^2 + v^2 + g^2)^{14}$$

of 16 factors, we must pick g from exactly 1 of the first 2 factors and we must pick a g^2 from 2 of the the last 14 factors; from each of the remaining 13 factors, we have 4 ways to pick a power of r, w, b, or v. The exponent 5 in g^5 is impossible to achieve by adding the multiples of 3 obtained from $(r^3 + w^3 + b^3 + v^3 + g^3)^{10}$. Finally, in the last term of the numerator, we must pick g^5 from exactly 1 of the 6 factors, and we must pick 1 of r^5, w^5, b^5, v^5 from each of the remaining 5 factors. Thus we have our computation

$$\frac{\big[\binom{30}{5}4^{25} + 15 \cdot \binom{2}{1}\binom{14}{2}4^{13} + 20 \cdot 0 + 24 \cdot \binom{6}{1}4^5\big]}{60},$$

which is 2,674,124,871,795,372,032, only slightly less humongous than the humongous number above.

Example. The octahedral group acting on the vertices of the cube has cycle index

$$\frac{z_1^8 + 9z_2^4 + 8z_1^2 z_3^2 + 6z_4^2}{24}.$$

So the number of 2-colorings of the vertices of the cube is

$$\frac{[1]2^8 + [9]2^4 + [8]2^2 2^2 + [6]2^2}{24},$$

or 24.

The pattern inventory of the black and white colorings of the vertices of a cube is

$$\frac{(b+w)^8 + 9(b^2+w^2)^4 + 8(b+w)^2(b^3+w^3)^2 + 6(b^4+w^4)^2}{24} =$$
$$b^8 w^0 + b^7 w^1 + 3b^6 w^2 + 3b^5 w^3 + 7b^4 w^4 + 3b^3 w^5 + 3b^2 w^6 + b^1 w^7 + b^0 w^8.$$

Homework.

1. How many ways are there to color the vertices of a cube so that there are 4 green vertices and 4 orange vertices? (Hint: No calculation is required.)

2. Find and then expand the pattern inventory under D_4 for the 2-colorings (black and white) of the 2-by-2 checkerboard.

3. Find and then expand the pattern inventory under the octahedral group for the 2-colorings (black and white) of the faces of a cube.

4. Find the pattern inventory under the octahedral group for the 3-colorings (red, white, and blue) of the edges of a cube.

5. How many ways under the octahedral group are there to color the edges of a cube such that 4 are red, 4 are white, and 4 are blue?

At first we might think that the next problem is misplaced; the problem should appear in the first half of the book where we were busy putting balls into boxes. Yet, the problem *is* here. In putting balls into boxes, we learned from experience that the word *indistinguishable* in front of the word *boxes* brings fear to the hearts of the bravest. However, if necessary, there is a technique that can help us solve such problems. We change the n given indistinguishable boxes to n distinguishable boxes and then use S_n acting on the distributions of balls into boxes to permute the boxes. In other words, any permutation of the boxes is considered to be the same distribution. Of course $|S_n|$ gets large with n and the problem is never trivial. We now give an example that uses this method.

Example. How many ways are there to put 50 black balls and 50 white balls into 4 indistinguishable boxes?

We begin with 4 boxes distinguished by the labels 1, 2, 3, 4. We let X be the set of all distributions of all the balls into the 4 labeled boxes. We want to be able to say that a distribution is "the same" as another if the second can be obtained from the first by permuting the labels on the boxes. This defines an equivalence relation, and we want to count the number of equivalence classes. (The permutations of the labels have the effect of making the boxes indistinguishable.) The group that is relevant here is the symmetric group S_4 of all permutations on $\{1, 2, 3, 4\}$. Specifically, if D is the distribution such that box i has b_i black balls and w_i white balls, for $i = 1, 2, 3, 4$, and if g is in S_4, then $\pi_g(D)$ is the distribution such that box $g(i)$ has b_i black balls and w_i white balls. The mapping $\pi : g \to \pi_g$ is an action of S_4 on X with the desired equivalence classes.

Table 5.4 does not explicitly give the cycle type for each element in the group S_4 but exhibits the cycle structure even more clearly so that we can more easily count the number of elements that have the same type. We should be able to give a simple argument for each entry in the middle

Element of S_4	# of that type	Black balls
$(\cdot)(\cdot)(\cdot)(\cdot)$	1	$\binom{4}{50}$
$(\cdot)(\cdot)(\cdot\cdot)$	$\binom{4}{2}$	26^2
$(\cdot)(\cdot\cdot\cdot)$	$4 \cdot 2$	17
$(\cdot\cdot)(\cdot\cdot)$	3	$\binom{2}{25}$
$(\cdot\cdot\cdot\cdot)$	3!	0

TABLE 5.4. Balls into 4 Distinguishable Boxes.

column of the table. For example, there are 3 permutations in S_4 that are the product of 2 transpositions, since such a permutation is determined by which of 2, 3, 4 is interchanged with 1. (Note that $\binom{4}{2}$ is not correct here, although $\binom{4}{2}/2$ is.) Since the sum of the middle column is 4!, we are confident that we have omitted none. Since the column entries for the white balls would be the same as those for the black balls in Table 5.4, the column for white balls is omitted. We now come to the third column of the table. We should be able to readily provide all the entries, except perhaps the second. In this case, if we put i black balls into each of the 2 boxes whose labels are interchanged, then we have left $50 - 2i$ black balls to put into the other 2 boxes. We can do this in $50 - 2i + 1$ ways. It then follows that we can distribute all 50 black balls in $51 + 49 + 47 + 45 + 43 + \cdots + 5 + 3 + 1$ ways. We want the sum of the first 26 odd integers. This is 26^2. In general,

$$\sum_{r=1}^{n}(2r - 1) = n^2$$

since $\sum_{r=1}^{n}(2r - 1) = 2\left[\sum_{r=1}^{n} r\right] - \left[\sum_{r=1}^{n} 1\right] = n(n + 1) - n = n^2$. Putting all the information from Table 5.4 together, we apply Burnside's Lemma to get the solution

$$\frac{\left\langle\begin{smallmatrix}4\\50\end{smallmatrix}\right\rangle^2 + 6 \cdot 26^4 + 8 \cdot 17^2 + 3 \cdot 26^2}{4!},$$

or $22,980,153$.

Homework for a Week.

1. Find the pattern inventory under the octahedral group for the 3-colorings (red, white, and blue) of the vertices of a cube.

2. How many ways can we 2-color (black and white) the 64 squares in the 8–by–8 checkerboard, with D_4 acting on the colorings, such that the number of white squares is greater than the number of black squares?

3. How many ways can the vertices of a cube be colored with 4 colors if each color is used twice?

4. For $k = 2$ and then for $k = 3$, find the pattern inventory for k-colorings ($b \& w$ or $r, w, \& b$) under D_n of the beads of an n-bead necklace where $n = 7$, $n = 9$, and $n = 11$. For each of these 6 cases, find the number of necklaces with exactly 3 white beads.

5. How many ways are there to 3-color a 4- by–4 checkerboard, under C_4, if each color is used at least once?

6. How many ways are there to put 12 indistinguishable balls into 3 indistinguishable boxes? How many ways are there to put 14 red balls, 4 white balls, and 4 blue balls into 3 indistinguishable boxes?

§41. Necklaces

Problems. How many ways are there to m-color the vertices of a regular n-gon under C_n? How many ways are there to m-color the vertices of a regular n-gon under D_n?

These problems are right up our alley. For small values of n, we can find the solution by Burnside's Lemma. For general values of n the procedure is the same but it gets a bit messy. If you have read §35, then we can answer the questions in general. Recall that if $d|n$ then C_n has $\phi(d)$ elements of order d. If α in C_n is of order d and if π is the natural action on the set of vertices of a regular n-gon, then π_α has n/d cycles of length d and so has cycle type $z_d^{n/d}$. When the problem involves D_n, there are 2 cases, depending on the parity of n. If n is odd, then every line of symmetry is a perpendicular bisector of a side of the n-gon. However, when n is even, then $n/2$ of the lines of symmetry are the perpendicular bisectors of the sides and $n/2$ are diagonals. Thus, we have the following as a corollary of Burnside's Lemma.

Theorem. The number of ways to k-color the vertices of a regular n-gon under C_n is

$$\frac{\sum_{d|n} \phi(d)k^{n/d}}{n}.$$

The number of ways to k-color the vertices of a regular n-gon under D_n is

$$\frac{\left[\sum_{d|n} \phi(d)k^{n/d}\right] + nk^{(n+1)/2}}{2n}$$

if n is odd and is

$$\frac{\left[\sum_{d|n} \phi(d)k^{n/d}\right] + \frac{n}{2}k^{n/2} + \frac{n}{2}k^{(n+2)/2}}{2n}$$

if n is even.

The nontrivial problems above are, of course, necklace problems. Now, we want to m-color the vertices of a regular n-gon with the restriction that no 2 adjacent vertices have the same color. In addition to Burnside's Lemma, we will need the following lemma. In the statement of the lemma, *fixed* means stationary, as opposed to *free*. (A free object can be moved about in space. For example, there are $\binom{6}{2}$ ways ways to color the faces of a fixed cube such that 2 are white and 4 are black, while there are only 2 ways to color the faces of a free cube such that 2 are white and 4 are black. The 2 white faces are either adjacent or else opposite.)

Lemma. The number of ways to m-color the vertices of a fixed regular n-gon such that no 2 adjacent vertices have the same color is

$$(m-1)^n + (-1)^n(m-1).$$

Proof. Since m will be constant throughout, we let $F(n)$ denote the number of ways to m-color the vertices of a fixed regular n-gon such that no 2 adjacent vertices have the same color. We easily see from Figure 5.11 that

$$F(3) = m(m-1)(m-2)$$

and that a fixed linear string of n beads can be m-colored such that no 2 adjacent beads have the same color in $m(m-1)^{n-1}$ ways.

From Figure 5.12, we deduce the equation

$$F(n) = m(m-1)^{n-1} - F(n-1)$$

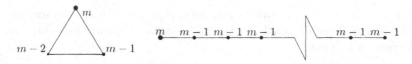

FIGURE 5.11. Triangles and Strings.

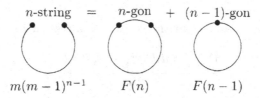

FIGURE 5.12. Necklaces and Strings.

for $n > 1$ as follows. We consider all the m-colorings of a fixed n-string. Those for which the ends have different colors are in one-to-one correspondence with the desired m-colorings of a fixed n-gon, while those m-colorings of the fixed n-string having the ends the same color are in one-to-one correspondence with the m-colorings of a fixed $(n-1)$-gon having no 2 adjacent colors the same.

A voice from the sky tells us to subtract $(m-1)^n$ from both sides of this equation and to go on to find a formula for $F(n)$. We could ask the voice for the formula and then, provided we get a response, prove that the given formula is correct simply by substituting the formula in the equation above, after checking that the formula is correct when $n = 3$. Not wishing to push it however, we are satisfied with the hint and do the following calculation.

$$
\begin{aligned}
F(n) - (m-1)^n &= m(m-1)^{n-1} - (m-1)^n - F(n-1) \\
&= (m-1)^{n-1} - F(n-1) \\
&= (-1)^1[F(n-1) - (m-1)^{n-1}] \\
&= (-1)^2[F(n-2) - (m-1)^{n-2}] \\
&= (-1)^3[F(n-3) - (m-1)^{n-3}] \\
&= (-1)^4[F(n-4) - (m-1)^{n-4}] \\
&\ \ \vdots \\
&= (-1)^{n-3}[F(3) - (m-1)^3] \\
&= (-1)^{n-3}[m(m-1)(m-2) - (m-1)^3] \\
&= (-1)^n(m-1),
\end{aligned}
$$

which proves the lemma. ∎

Example. How many ways can an 8-bead necklace be 3-colored under D_8 such that no 2 adjacent beads have the same color?

We are 3-coloring the vertices of a regular octagon under D_8 such that no 2 adjacent vertices have the same color. For each element g in D_8, we will compute $|F_g^\star|$ and then use Burnside's Lemma. We will need the lemma directly above more than once in the computation. First, that $|F_i^\star| = 2^8 + 2$ comes directly from the lemma with $m = 3$ and $n = 8$.

We next look at $|F_{\rho^4}^\star|$. Now, ρ^4 is a rotation of $180°$. Let $A = (1,0)$, $B = (1/\sqrt{2}, 1/\sqrt{2})$, $C = (0,1)$, $D = (-1/\sqrt{2}, 1/\sqrt{2})$, and $E = (-1,0)$. The coloring of A, B, C, D determines the colors of the remaining vertices in this case and, since π_{ρ^4} takes A to E, then A and E must then be the same color. So we identify A and E to form the "loop" $ABCD$, which we may as well consider a fixed 4-bead necklace. Thus, $|F_{\rho^4}^\star| = 2^4 + 2$ from the lemma with $m = 3$ and $n = 4$. Likewise, each of the fix of ρ^2 and the fix of ρ^6 has $2^2 + 2$ elements by a similar application of the lemma or else is seen to be $3 \cdot 2$ by direct counting. Of course, the number of elements in the fix of each of ρ, ρ^3, ρ^5, ρ^7 is 0.

By counting the number of 3-colorings of a 5-string, we see that the number of elements in the fix of each of the reflections σ_h, σ_v, σ_p, σ_m is $3 \cdot 2^4$. For each of the other 4 reflections, which are reflections in the perpendicular bisectors of the sides of the octagon, the number of elements in the fix is necessarily 0.

Our final answer to the posed question is, by Burnside's Lemma,

$$\frac{[1](2^8 + 2) + [1](2^4 + 2) + [2](2^2 + 2) + [4]0 + [4](3 \cdot 2^4) + [4]0}{16},$$

or 30.

Exercises.

1. How many 4-bead necklaces made from ruby red, diamond white, and emerald green beads (use r, w, g) are there if no 2 adjacent beads can have the same color?

2. How many 28-bead necklaces can be made from beads of 3 colors with the restriction that no 2 adjacent beads have the same color?

3. How many ways are there to 5-color the 9 vertices in the figure below provided that no 2 adjacent (connected) vertices can have the same color? Assume that the figure is free to rotate and turn over.

4. How many ways are there to 5-color the 9 vertices in the figure below provided that no 2 adjacent (connected) vertices can have the same color? Assume that the figure is free to rotate and turn over.

Practice Exam.

Before you begin, you should have a cut-out equilateral triangle that you can fold (along the 3 joins of the midpoints of the sides) to produce a tetrahedron that is sufficient to help determine the **tetrahedral group**, the group of all rotation symmetries of the regular tetrahedron.

1. Under the tetrahedral group, how many ways are there to m-color the faces of a regular tetrahedron?

2. Under the tetrahedral group, how many ways are there to m-color the vertices of a regular tetrahedron?

3. Under the icosahedral group, how many ways are there to color the faces of a regular dodecahedron such that there are 5 red faces, 2 white faces, and 5 blue faces?

4. Under the dihedral group D_{12}, how many ways are there to 5-color a 12 bead necklace such that no 2 adjacent beads have the same color?

6

Recurrence Relations

§42. Examples of Recurrence Relations

A **recurrence relation** for a sequence $\{a_i\}$, which usually begins with a_0 or a_1, is a formula that defines a_n in terms of $a_0, a_1, a_2, a_3, \ldots, a_{n-1}$, and n for all n greater than some particular integer k, with the terms $a_0, a_1, \ldots,$ a_k called **initial conditions** or **boundary conditions**. Together with the initial conditions, the recurrence relation provides a recursive definition for the elements of the sequence. This allows us to compute the unique value of a_n for each integer n such that $n > k$. Many examples follow. We are already familiar with several recurrence relations.

Example 1. *Factorials.* Let $a_0 = 1$ and $a_n = na_{n-1}$ for $n > 0$. We immediately recognize this recurrence relation with its initial condition as the recursive definition of $n!$ We have $a_n = n!$ for integer n such that $n \geq 0$.

Example 2. *Interest.* Let r denote the interest rate paid for the time period of compounding; let P be the initial payment; and let A_n be the amount accumulated after n time periods. We have $A_n = A_{n-1} + rA_{n-1}$ with $A_0 = P$. This, of course, gives $A_n = P(1 + r)^n$ for all n. Much of the business of the world is based on the application of compound interest.

Example 3. *A chess story.* Let $a_n = 2a_{n-1}$ for $n > 1$ with $a_1 = 1$. This comes with a story about the invention of the game of chess and the number $18,446,744,073,709,551,615$, which is $2^{64} - 1$. Legend has it that the inventor of chess was rewarded by the king for this marvelous game by having any wish granted. The king was at first pleased that the inventor merely asked for 1 grain of wheat for the first square of the chess board, 2 grains of wheat for the second square, 4 grains for the third, and each successive square was to have twice the number of grains as the previous square. However, when it became evident that there has never been enough wheat grown throughout history to come even close to accomplishing this, the king simply beheaded the inventor. As in Examples 1 and 2, the formula $a_n = 2^{n-1}$ for a_n is obtained by "unwinding." (Actually, Example 3 is Example 2 with an astonishing interest rate.)

This unwinding in Example 3 might proceed from the bottom up as follows: $a_1 = 1, a_2 = 2a_1 = 2, a_3 = 2a_2 = 2^2, a_4 = 2a_3 = 2^3, a_5 = 2a_4 = 2^4$, $a_6 = 2a_5 = 2^5, \ldots, a_n = 2a_{n-1} = 2^{n-1}$. An alternative unwinding might proceed from the top down as follows: $a_n = 2a_{n-1} = 2 \cdot 2a_{n-2} = 2^2 a_{n-2} = 2^2 \cdot 2a_{n-3} = 2^3 a_{n-3} = 2^3 \cdot 2a_{n-4} = 2^4 a_{n-4} = 2^4 \cdot 2a_{n-5} = 2^5 a_{n-5} = \cdots = 2^{n-2} a_{n-(n-2)} = 2^{n-2} a_2 = 2^{n-2} \cdot 2a_1 = 2^{n-1} a_1 = 2^{n-1}$.

Example 4. *Bell numbers.* The word "partition" is used in more than 1 way in mathematics. Earlier, in Chapter 2, we talked about the number $\Pi(r)$ of partitions of a *positive integer* r. Now, we want to talk about the number of partitions of a *set* S. A **partition** of a set S is a set of disjoint nonempty subsets of S whose union is S. So, every element of S is in exactly 1 of the subsets. In particular, we want to count the number of partitions of the set $\{1, 2, 3, 4, \ldots, n\}$. We will denote this number by B_n. These numbers are called **Bell numbers**, after Eric Temple Bell (1883–1960). Every mathematician has read Bell's *Men of Mathematics*, a history of mathematics that may not always be accurate but which never fails to be entertaining. For example, $B_3 = 5$:

$$\{1, 2, 3\}, \quad \{1, 2\} \cup \{3\}, \quad \{1, 3\} \cup \{2\}, \quad \{2, 3\} \cup \{1\}, \quad \{1\} \cup \{2\} \cup \{3\}.$$

The number B_n also counts the number of different equivalence relations that can be defined on a set of n elements. (Elements in the set are equivalent iff they are in the same subset.) We take $B_0 = B_1 = 1$ and derive a recurrence relation for calculating B_{n+1} for $n > 0$ as follows. In a partition of the first $n + 1$ positive integers, the number $n + 1$ lies in some subset of size $k + 1$ where $0 \le k \le n$. There are $\binom{n}{k}$ choices for the k elements that are in the same subset as $n + 1$, and the remaining $(n + 1) - (k + 1)$ elements can be partitioned into subsets in $B_{(n+1)-(k+1)}$ ways. So

$$B_{n+1} = \sum_{k=0}^{n} \binom{n}{k} B_{(n+1)-(k+1)}$$

$$= \sum_{k=0}^{n} \binom{n}{k} B_{n-k} = \sum_{k=0}^{n} \binom{n}{n-k} B_{n-k}$$

$$= \sum_{k=0}^{n} \binom{n}{k} B_k.$$

Thus the Bell numbers are determined by the recurrence relation

$$B_0 = 1 \quad \text{and} \quad B_{n+1} = \sum_{k=0}^{n} \binom{n}{k} B_k \text{ for } n \geq 0.$$

The Bell numbers are actually not that new to us. How many ways are there to put n distinguishable balls into n indistinguishable boxes?

$$B_n = \sum_{k=1}^{n} \left\{ {n \atop k} \right\}$$

for $n > 0$, where $\left\{ {n \atop k} \right\}$ is a Stirling number of the second kind. Also,

$$B_n = \frac{1}{e} \sum_{k=0}^{\infty} \frac{k^n}{k!}.$$

Example 5. *Derangements.*

$$D_n = nD_{n-1} + (-1)^n \text{ for } n > 1 \text{ with } D_1 = 0.$$

Example 6. *Subfactorials.* We also know that

$$D_{n+1} = n[D_n + D_{n-1}] \text{ for } n > 2 \text{ with } D_1 = 0 \text{ and } D_2 = 1.$$

Note that $a_{n+1} = n[a_n + a_{n-1}]$, the same recurrence relation as that for D_n, but with the boundary conditions $a_0 = a_1 = 1$, is not the sequence $\{D_n\}$ shifted but rather the more familiar sequence $\{n!\}$. Thus, using the subfactorial notation n_i for D_n, we see that both n_i and $n!$ satisfy this recurrence relation but, necessarily, with different initial conditions.

The definition of a recurrence relation can be generalized to include arrays $\{a_{n,k}\}$, as in the next 3 examples.

Example 7. *Binomial coefficients.* Let $a_{n,k} = a_{n-1,k-1} + a_{n-1,k}$ for $0 < k < n$, with boundary conditions $a_{n,0} = a_{n,n} = 1$. This recurrence relation gives $a_{n,k} = \binom{n}{k}$ since the recurrence relation is the familiar relation

$$\binom{n}{k} = \binom{n-1}{k-1} + \binom{n-1}{k},$$

which, together with the boundary conditions $\binom{n}{0} = \binom{n}{n} = 1$ for $n \geq 0$, defines Pascal's Triangle.

Example 8. *Partition of an integer.* For positive integers r and n, define $\Pi(r,n)$ by $\Pi(r,r) = \Pi(r,1) = 1$, $\Pi(r,n) = 0$ if $n > r$, and otherwise by

$$\Pi(r,n) = \Pi(r-1, n-1) + \Pi(r-n, n).$$

We do not know a closed form for $\Pi(r, n)$, the number of ways to partition integer r into n parts, which is also the number of ways to put r indistinguishable balls into n indistinguishable boxes with no box empty.

Example 9. *Stirling numbers.* Define $\left\{ {r \atop n} \right\}$ by $\left\{ {r \atop r} \right\} = \left\{ {r \atop 1} \right\} = 1$, $\left\{ {r \atop n} \right\} = 0$ if $n > r$, and otherwise by

$$\left\{ {r \atop n} \right\} = \left\{ {r - 1 \atop n - 1} \right\} + n \left\{ {r - 1 \atop n} \right\}.$$

We do know a closed form for $\left\{ {r \atop n} \right\}$, the number of ways to put r distinguishable balls into n indistinguishable boxes with no box empty.

Example 10. *Unwinding.* Suppose a_n is defined by $a_n = a_{n-1} + n$ for $n > 1$ with $a_1 = 1$. We again illustrate the important technique of "unwinding." The bottom-up attack proceeds as follows:

$$a_1 = 1,$$
$$a_2 = a_1 + 2 = (1) + 2 = 1 + 2,$$
$$a_3 = a_2 + 3 = (1 + 2) + 3 = 1 + 2 + 3,$$
$$a_4 = a_3 + 4 = (1 + 2 + 3) + 4 = 1 + 2 + 3 + 4,$$
$$\vdots$$
$$a_n = a_{n-1} + n = (1 + 2 + 3 + \cdots + (n - 1)) + n = 1 + 2 + 3 + \cdots + n.$$

So,

$$a_n = \sum_{k=1}^{n} k = \binom{n + 1}{2} = \frac{n(n + 1)}{2}.$$

The top-down attack proceeds as follows:

$$a_n = a_{n-1} + n = [a_{n-2} + (n - 1)] + n$$
$$= a_{n-2} + [(n - 1) + n] = [a_{n-3} + (n - 2)] + [(n - 1) + n]$$
$$= a_{n-3} + [(n - 2) + (n - 1) + n]$$
$$= [a_{n-4} + (n - 3)] + [(n - 2) + (n - 1) + n]$$
$$= a_{n-4} + [(n - 3) + (n - 2) + (n - 1) + n]$$
$$\vdots$$
$$= a_{n-(n-1)} + [(n - (n - 2)) + \cdots + (n - 2) + (n - 1) + n]$$
$$= a_1 + [2 + 3 + 4 + \cdots + (n - 2) + (n - 1) + n]$$
$$= 1 + 2 + 3 + 4 + \cdots + (n - 2) + (n - 1) + n,$$

as before.

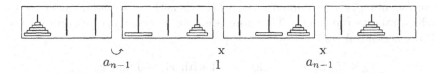

$$a_{n-1} \qquad \begin{matrix} \text{x} \\ 1 \end{matrix} \qquad \begin{matrix} \text{x} \\ a_{n-1} \end{matrix}$$

FIGURE 6.1. Tower of Hanoi.

Example 11. The *Tower of Hanoi*. We have another story about $2^{64} - 1$, this time concerning the end of the world. In the beginning, God placed 64 gold disks of different diameters on the first of 3 diamond needles so that each disk was above only larger ones and ordered the priests of the Tower of Brahma to move them to the third needle with the stipulation that a larger disk was never to be placed on top of a smaller disk and that only 1 disk could be moved to 1 of the needles at a time. When the priests finish this task the Tower of Brahma will crumble and the world will end. This tale accompanied a toy puzzle called the Tower of Hanoi, which was invented by the French mathematician Édouard Lucas in 1883. The original toy had 8 disks. The general problem is to find the minimum number of moves required to complete the task if we start with n disks.

From Figure 6.1, it is seen that the recurrence relation for the Tower of Hanoi is $a_n = a_{n-1} + 1 + a_{n-1} = 2a_{n-1} + 1$. We may take $a_0 = 0$ (or $a_1 = 1$). Unwinding, we get $a_n = 2a_{n-1} + 1 = 2[2a_{n-2} + 1] + 1 = 2^2 a_{n-2} + 2^1 + 2^0 = 2^2[2a_{n-3} + 1] + 2^1 + 2^0 = 2^3 a_{n-3} + 2^2 + 2^1 + 2^0 = \cdots = 2^n a_{n-n} + 2^{n-1} + 2^{n-2} + \cdots + 2^2 + 2^1 + 2^0 = 2^n - 1$.

Exercise.

Unwind the recurrence relation $a_n = 2a_{n-1} + 2$ with the boundary condition $a_0 = 0$ both from the bottom up and from the top down.

§43.   The Fibonacci Numbers

Example 12. The *Fibonacci numbers*. The Fibonacci recurrence relation F_n is the granddaddy of them all. In 1202 Leonardo of Pisa, who is known as Fibonacci, posed the problem below in his book *Liber Abaci*. The story this time concerns rabbits. We suppose that a pair of adult rabbits produces 1 pair of young rabbits of opposite sex each month, that newborn rabbits produce their first offspring at the end of 2 months, and that rabbits never die. We begin with 1 newborn pair of rabbits at the beginning of month 1, having 0 pairs of rabbits at the beginning of month 0, and let F_k denote the number of pairs of rabbits on hand at the beginning of the kth month. So $F_0 = 0$ and $F_1 = 1$. At the beginning of any month after that, we have all the pairs of rabbits that were alive at the beginning of last month, since rabbits never die, and the newborn pairs whose number equals the number of pairs alive 2 month ago, since this is the amazing frequency with which rabbits reproduce. Therefore, we have a recurrence relation for the Fibonacci numbers:

$$F_n = F_{n-1} + F_{n-2} \text{ for } n > 1 \text{ with } F_0 = 0 \text{ and } F_1 = 1.$$

The Fibonacci sequence begins

$$0, 1, 1, 2, 3, 5, 8, 13, 21, 34, 55, 89, 144, 233, 377, 610, 987, 1597, \ldots$$

We note that the first and sometimes even the second of the terms above are omitted from the sequence. There is much written about this sequence and its generalizations. The interesting sequence $\{G_n\}$, defined by $G_n = F_{n+1}/F_n$ for $n > 0$, describes the growth rate of the Fibonacci sequence and begins: $1, 2, 1.5, 1.66\ldots, 1.6, 1.625, 1.615\ldots, 1.619\ldots, 1.617\ldots,$ $1.6181\ldots, 1.6179\ldots, 1.61805\ldots, 1.61802\ldots, 1.618037\ldots, 1.618032\ldots,$ $1.618034\ldots, 1.6180338\ldots$. It looks like this sequence approaches some constant g. Assuming that this is the case, we can easily find g. Since

$$G_n = \frac{F_{n+1}}{F_n} = \frac{F_n + F_{n-1}}{F_n} = 1 + \frac{F_{n-1}}{F_n} = 1 + \frac{1}{G_{n-1}}.$$

in the limit we have $g = 1 + 1/g$ or $g^2 - g - 1 = 0$. So

$$g = \frac{1 + \sqrt{5}}{2}.$$

A calculator approximates *the golden ratio* g as 1.6180339. Note that we have not proved that the sequence $\{G_n\}$ actually approaches g. We have proved that if the sequence $\{G_n\}$ does have a limit, then this limit must be g. We do not yet have a formula for F_n, which would settle the matter. However, it appears that, when we find a formula, such a formula must be somewhat complicated in order to have F_{n+1}/F_n approach $(1 + \sqrt{5})/2$. What has $\sqrt{5}$ to do with rabbits? We could prove the correctness of the formula if it were presented now, but we prefer to derive the formula later.

We can easily find the generating function $F(z)$ for the Fibonacci sequence. Subtracting the second and third equations from the first in

$$F(z) = F_0 + F_1 z + F_2 z^2 + F_3 z^3 + F_4 z^4 + F_5 z^5 + \cdots,$$
$$zF(z) = \quad\quad F_0 z + F_1 z^2 + F_2 z^3 + F_3 z^4 + F_4 z^5 + \cdots,$$
$$z^2 F(z) = \quad\quad\quad\quad F_0 z^2 + F_1 z^3 + F_2 z^4 + F_3 z^5 + \cdots,$$

we get $F(z) - zF(z) - z^2 F(z) = F_0 + (F_1 - F_0)z = z$. Thus,

$$F(z) = \frac{z}{1 - z - z^2}.$$

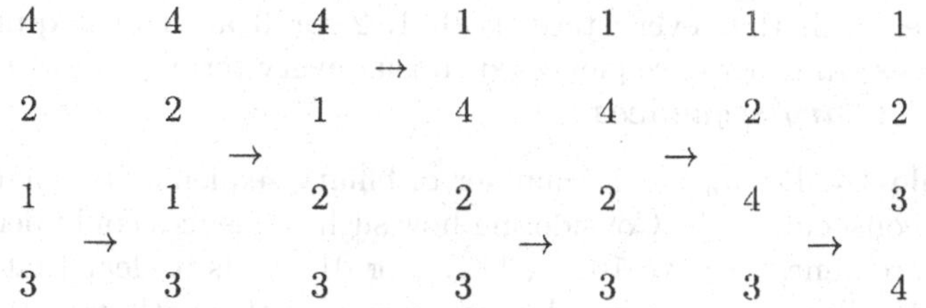

4	4	4	1	1	1	1
2	2	1	4	4	2	2
1	1	2	2	2	4	3
3	3	3	3	3	3	4

FIGURE 6.2. Bubble Sort.

MORE PIE PROBLEMS? (This time it is a Pizza Pie!)

Without what follows, we would probably not think to make the observation that 0 planes separate space into 1 region. However, we can go on to say that 1 plane separates space into 2 regions; 2 planes separate space into at most 4 regions; and 3 planes separate space into at most 8 regions. Into at most how many regions do 4 planes separate space? Model the 3 following related problems with a recurrence relation and solve each with a formula. A version of the second has been called the Pizza Problem, as instead of the plane the problem is to cut up a pizza. Hint: Do the problems in reverse order, using each to help solve the next.

Space Pizza Find a_n, the maximum number of 3-dimensional regions in the separation of space by n planes.

Plane Pizza Find b_n, the maximum number of 2-dimensional regions in the separation of the plane by n lines.

Line Pizza Find c_n, the maximum number of 1-dimensional regions in the separation of a line by n points.

Example 13. *Bubble sort.* Given a list of n numbers, we wish to place them in order. Starting at the bottom of the list we successively compare each item with the number above it, interchanging the pair iff the bottom one is smaller than the top one. The smallest number on our list has bubbled to the top after $n - 1$ comparisons. We are then faced with the same problem with a shorter list of $n - 1$ numbers. Hence, we have a recurrence relation for a_n, the number of necessary comparisons: $a_n = a_{n-1} + (n - 1)$ for $n \geq 1$ with $a_1 = 0$. For example, $a_4 = 6$, as we see from Figure 6.2, where we start with the left column and progress to the right column making interchanges as necessary. It follows by unwinding that $a_n = a_1 + 1 + 2 + \cdots + (n - 2) + (n - 1) = n(n - 1)/2$.

Sequences such that every term is 0 or 1 are called **binary sequences**. Sequences such that every term is 0, 1, or 2 are called **ternary sequences**.

Sequences such that every term is 0, 1, 2, or 3 are called **quaternary sequences**. In general, sequences such that every term is one of n symbols are called *n-ary sequences*.

Example 14. Let a_n be the number of binary sequences of length n that have no consecutive 0's. Considering how such sequences could begin under our requirement, we have $10...$, $11...$, or $01....$ Is it clear that we must have a 1 followed by such a binary sequence of length $n-1$ or else a 01 followed by such a binary sequence of length $n-2$? Conversely, every such binary sequence constructed in this fashion satisfies our requirements. So $a_n = a_{n-1} + a_{n-2}$. We have seen this recurrence relation before. Our boundary conditions here are $a_1 = 2$ and $a_2 = 3$. This gives the Fibonacci sequence with a couple terms chopped off. Therefore, we have the somewhat surprising result that $a_n = F_{n+2}$ for $n \geq 0$.

We could have used $a_0 = 1$ and $a_1 = 2$ as the boundary conditions above. This necessitates introducing the sequence of length 0. This **empty sequence** is traditionally denoted by λ and can be very handy. After all, if we can have a set with no elements, why can we not have a sequence with no terms?

Example 15. How many ternary sequences are there with no 0's appearing to the right of some 2? We are looking at sequences that have the form

$$< \text{anything} > \quad 2 \quad < \text{no 0's} >.$$

We will model the problem with a recurrence relation in 3 different ways.

We first consider the beginning of these sequences. Suppose that there are b_n such sequences. There are b_{n-1} of these that begin with a 0 and the same number that begin with a 1. In counting these sequences that begin with a 2, we see that this 2 can be followed by any such sequence of length $n-1$ or else by the sequence of all 1's. We have the figure

$$
\begin{array}{ll}
0.......... & b_{n-1} \\
1.......... & b_{n-1} \\
2.......... & b_{n-1} + 1
\end{array}
$$

So $b_n = 3b_{n-1} + 1$ with $b_1 = 1$.

Now we model the same problem but this time consider the end of these sequences. We suppose that there are c_n such sequences. None of these sequences can end with a 0. There are c_{n-1} that end with a 1. Finally, there are 3^{n-1} that end with a 2. Here, we have the figure

$$
\begin{array}{ll}
..........0 & 0 \\
..........1 & c_{n-1} \\
..........2 & 3^{n-1}
\end{array}
$$

So $c_n = c_{n-1} + 3^{n-1}$ with $c_1 = 1$.

Of course the recurrence relation $b_n = 3b_{n-1} + 1$ with $b_1 = 1$ and the recurrence relation $c_n = c_{n-1} + 3^{n-1}$ with $c_1 = 1$ must necessarily have the same solution. They do; we can check that

$$b_n = c_n = \frac{3^n - 1}{2} \text{ for all } n$$

by substituting in each recurrence relation and also checking that the formula gives the correct initial conditions. For example, in order to check for the second recurrence relation we need to verify that

$$\frac{3^n - 1}{2} = \frac{3^{n-1} - 1}{2} + 3^{n-1} \text{ and } \frac{3^1 - 1}{2} = 1.$$

We will soon learn how to find such solutions, at least in nice cases. In this example, however, by considering what comes before and after the last 2, we might be able to see immediately that $a_n = \sum_{k=0}^{n-1} 3^k = (3^n - 1)/2$ for $n \geq 0$.

§44.   A Dozen Recurrence Problems

The Dozen. Model each of the following problems with a recurrence relation and its initial conditions.

1. Let a_n be the number of regions formed in the plane by n mutually overlapping circles, no 3 intersecting at a common point.

2. Cars are parked in line as they come off 3 different assembly lines, ready for later road testing. Today's production produces 3 models, each of a single color. Each red car takes 2 spaces and each blue car takes 2 spaces, while the green cars take only 1 space each. Let a_n be the number of ways of filling the first n parking spaces.

3. We can climb n stairs in a_n ways where each step we take consists of either 1 stair or 2 stairs and the order of the steps is considered.

4. Let a_n be the number of ternary sequences of length n that have no consecutive 0's.

5. Let a_n be the number of subsets of $\{1, 2, 3, \ldots, n\}$ that contain no consecutive integers.

6. Let a_n be the number of ways to arrange n feet of flags from the top of a flagpole if the flags available are 1-foot red flags, 2-foot white flags, and 1-foot blue flags.

7. This year's change in a_n is always twice last year's change, where $a_0 = 2$ and $a_1 = 7$.

8. Where the order of purchase is considered, let a_n be the number of ways to spend n dollars if for \$1 we can buy either a red ball or a white ball and for \$2 we can buy either a blue, green, or black ball.

9. Let a_n be the number of ways to stack n poker chips, each of which is red, white, blue, or green, with no consecutive green chips.

10. Let a_n be the number of ways a coin can be flipped n times such that the second head appears on or before the nth flip.

11. Let a_n be the number of comparisons that must be made between pairs of numbers in order to determine the maximum and minimum of a set of 2^n distinct real numbers.

12. Let $a_{r,n}$ be the number of ways to distribute r balls into n distinguishable boxes with 2, 3, or 4 balls in each box if there are at least $4n$ red balls, $4n$ white balls, and $4n$ blue balls available.

§45. Solving Recurrence Relations

Suppose p is a positive integer and we have the recurrence relation

$$a_n = c_1 a_{n-1} + c_2 a_{n-2} + c_3 a_{n-3} + \cdots + c_{p-1} a_{n-p+1} + c_p a_{n-p}$$

for $n \geq p$, with given constants c_i. We shall call this recurrence relation $(\star)$ for convenience. We suppose $c_p \neq 0$. If we replace a_k by x^k throughout $(\star)$, we have the polynomial equation

$$x^n = c_1 x^{n-1} + c_2 x^{n-2} + c_3 x^{n-3} + \cdots + c_{p-1} x^{n-p+1} + c_p x^{n-p}.$$

Now, it is easy to see that if r is one of the roots of this polynomial equation, then $a_n = r^n$ is a solution to $(\star)$. We factor out the $n - p$ zero roots of the polynomial by dividing both sides of the equation by x^{n-p} to get

$$x^p = c_1 x^{p-1} + c_2 x^{p-2} + c_3 x^{p-3} + \cdots + c_{p-1} x + c_p,$$

which is called the **characteristic equation** of $(\star)$.

We now show that a linear combination of solutions to $(\star)$ is a solution to $(\star)$. Specifically, we want to show that if $a_n = a'_n$ and $a_n = a''_n$ are solutions to $(\star)$, then $a_n = b_1 a'_n + b_2 a''_n$ is a solution to $(\star)$ for arbitrary constants b_1 and b_2. So, assuming

$$a'_n = c_1 a'_{n-1} + c_2 a'_{n-2} + \cdots + c_p a'_{n-p}$$

and

$$a_n'' = c_1 a_{n-1}'' + c_2 a_{n-2}'' + \cdots + c_p a_{n-p}'',$$

then

$$b_1 a_n' + b_2 a_n'' =$$
$$c_1 [b_1 a_{n-1}' + b_2 a_{n-1}''] + c_2 [b_1 a_{n-2}' + b_2 a_{n-2}''] + \cdots + c_p [b_1 a_{n-p}' + b_2 a_{n-p}'']$$

follows by multiplying both sides of the first equation by b_1, multiplying both sides of the second equation by b_2, and then adding the columns. So $a_n = b_1 a_n' + b_2 a_n''$ is a solution to $(\star)$, as desired.

Applying this to our comments about the roots of the characteristic equation of $(*)$, we have an important result.

Observation. If the characteristic equation of the recurrence relation

$$a_n = c_1 a_{n-1} + c_2 a_{n-2} + c_3 a_{n-3} + \cdots + c_{p-1} a_{n-p+1} + c_p a_{n-p},$$

for $n \geq p$, has nonzero roots $r_1, r_2, \ldots, r_p$, then the recurrence relation has solutions

$$a_n = b_1 r_1^n + b_2 r_2^n + b_3 r_3^n + \cdots + b_p r_p^n$$

where the b_i are arbitrary constants.

The p boundary conditions that are necessary to determine the a_n for all n given by $(\star)$ will uniquely determine the constants b_i, provided that the roots are all different. In this case, we have p equations with p unknowns. It is not altogether obvious that there will always be a unique solution. Although this will be the case, we need not worry about it because we know there is a unique solution to the recurrence relation together with its boundary conditions and this process always provides us with a solution, which must necessarily then be the unique solution. A couple examples will illustrate the process.

Example 16. Suppose $a_n = 2a_{n-1} - 2a_{n-2}$ for $n \geq 2$, with $a_0 = 0$ and $a_1 = -2$. Here, $p = 2$. We replace a_k by x^k in the recurrence to get the polynomial equation $x^n = 2x^{n-1} - 2x^{n-2}$. Dividing both sides of the equation by x^{n-2}, we get the characteristic equation $x^2 = 2x - 2$. This has roots $x = 1 \pm i$, where $i^2 = -1$. So our recurrence has general solution $a_n = b(1+i)^n + c(1-i)^n$ with arbitrary constants b and c. Now, from the boundary conditions we have the 2 equations $0 = a_0 = b(1+i)^0 + c(1-i)^0$ and $-2 = a_1 = b(1+i)^1 + c(1-i)^1$. That is,

$$0 = b + c,$$
$$-2 = b(1+i) + c(1-i),$$

which has the unique solution $b = i$, $c = -i$. Hence, we have our final formula for our given recurrence relation with its boundary conditions:

$$a_n = i(1+i)^n - i(1-i)^n \text{ for } n \geq 0.$$

We may be somewhat surprised that complex numbers have come into play in dealing with a recurrence relation that has only integer values.

Example 17. We return to the Fibonacci recurrence relation

$$F_n = F_{n-1} + F_{n-2} \text{ for } n \geq 2 \text{ with } F_0 = 0 \text{ and } F_1 = 1.$$

Here, we get the characteristic equation $x^2 = x + 1$, which has the roots $(1 \pm \sqrt{5})/2$. So $F_n = a[(1+\sqrt{5})/2]^n + b[(1-\sqrt{5})/2]^n$ for arbitrary constants a and b is a general solution to the recurrence $F_n = F_{n-1} + F_{n-2}$. The boundary conditions give the equations $0 = F_0 = a + b$ and $1 = F_1 = a(1+\sqrt{5})/2 + b(1-\sqrt{5})/2$. These have the solution $a = 1/\sqrt{5}$, $b = -1/\sqrt{5}$ and we have our desired formula for the Fibonacci sequence:

$$F_n = \frac{\left[\frac{1+\sqrt{5}}{2}\right]^n - \left[\frac{1-\sqrt{5}}{2}\right]^n}{\sqrt{5}} \text{ for } n \geq 0.$$

Needless to say, it is easier to calculate F_{20} in standard decimal notation by using the recurrence relation than by substituting 20 in the formula above.

What happens if the roots to the characteristic equation are not distinct? That is, what happens when the characteristic equation has multiple roots? Suppose r is a root of multiplicity m. This means that exactly m of the roots are r, which is equivalent to saying that m is the highest power of $(x-r)$ that divides the characteristic equation. We suppose $m > 1$. In this case,

$$d_1 r^n + d_2 n r^n + d_3 n^2 r^n + \cdots + d_m n^{m-1} r^n \text{ with arbitrary } d_i$$

replaces the redundant $d_1 r^n + d_2 r^n + \cdots + d_m r^n$ in our Observation above. Again, this will always work if all multiple roots are handled in this manner. (Students of calculus will understand that if r is a double root of the characteristic equation, then r is a root of the derivative and, so, nr^n is a solution to the recurrence. With a triple root, we can play this game again and get that $n^2 r^n$ is another solution, and so on. See the following example.)

Example 18. Consider $a_n = 3a_{n-1} - 4a_{n-3}$ for $n \geq 3$ with $a_0 = 0$, $a_1 = 2$, $a_2 = -1$. The characteristic equation $x^3 = 3x^2 - 4$ has roots 2, 2, -1. In other words, $x^3 - 3x^2 + 4 = (x-2)^2(x+1)$. Not only is $a_n = 2^n$ a solution to the recurrence but $a_n = n2^n$ is also a solution; we know $a_n = (-1)^n$ is another solution to the recurrence. (For calculus students only: From "$x^n = 3x^{n-1} - 4x^{n-3}$ has double root 2" to "$nx^{n-1} = 3(n-1)x^{n-2} - 4(n-3)x^{n-4}$ has root 2" by calculus to "$[nx^n] = 3[(n-1)x^{n-1}] - 4[(n-3)x^{n-3}]$ has root 2" by simple algebra to "$[n2^n] = 3[(n-1)2^{n-1}] - 4[(n-3)2^{n-3}]$" by simple algebra to "$a_n = n2^n$ is a solution to $a_n = 3a_{n-1} - 4a_{n-3}$.") We have the general solution

$$a_n = a2^n + bn2^n + c(-1)^n \text{ with } a, b, c \text{ arbitrary.}$$

We take the boundary conditions into consideration to get

$$0 = a_0 = a \qquad + c,$$
$$2 = a_1 = 2a + 2b - c,$$
$$-1 = a_2 = 4a + 8b + c.$$

Solving these 3 equations, we get $a = 1$, $b = -1/2$, and $c = -1$. Therefore, we have the formula for our recurrence

$$a_n = 2^n - n2^{n-1} + (-1)^{n+1} = 2^{n-1}(2 - n) + (-1)^{n+1} \text{ for } n \geq 0.$$

Mathematical limitations on solving polynomial equations imply that, in general, it is impossible to find a closed formula to solve a recurrence relation. Nevertheless, in the age of the computer, modeling by recurrence relations is much more that an academic exercise. Computers are a natural when it comes to many recursive computations to get at a desired particular solution.

Homework.

Solve, with a closed formula, the first 8 of the dozen problems of the previous section.

§46. The Catalan Numbers

Example 19. How many ways can $n + 1$ numbers be multiplied together? Let's be more precise. Let C_n be the number of ways we can insert parentheses in the product

$$x_1 * x_2 * x_3 * x_4 * x_5 * x_6 * x_7 * \cdots * x_n * x_{n+1}$$

so that we have n successive multiplications, each involving 2 factors. Because of the associative law, the product is the same for each final product. So the question is really how many ways can we associate a product of $n + 1$ numbers, with the numbers in a given order. Since division is not associative, C_n is also the answer to the question, How many ways can the $n + 1$ nonzero numbers x_i in the otherwise meaningless form

$$x_1 \div x_2 \div x_3 \div x_4 \div x_5 \div x_6 \div x_7 \div \cdots \div x_n \div x_{n+1}$$

be grouped so that the n division signs give a well-defined number?

For example, $C_2 = 2$ from $a * (b * c)$ and $(a * b) * c$, and $C_3 = 5$ from

$$a * (b * (c * d)), \quad a * ((b * c) * d), \quad (a * (b * c)) * d,$$
$$((a * b) * c) * d, \quad \text{and} \quad (a * b) * (c * d).$$

Now, $C_1 = 1$ and we accept $C_0 = 1$. Let's look at a computation for C_{n+1}. Since the last of the $n + 1$ multiplications is a product of 2 factors such that the first factor involves the first $i + 1$ of the $n + 2$ numbers, which can be associated in C_i ways, and the second factor involves the remaining $n + 2 - (i + 1)$, or $(n - i) + 1$, numbers, which can be associated in C_{n-i} ways, we have the recurrence relation

$$C_{n+1} = C_0 C_n + C_1 C_{n-1} + C_2 C_{n-2} + C_3 C_{n-3} + \cdots + C_{n-1} C_1 + C_n C_0$$

for $n > 0$ with $C_0 = 1$.

This is a solution, but we can do better if we also consider a related problem. Let A_n be the answer to the same problem as above, except this time allowing ourselves the possibility of reordering the given numbers in the product. So A_n is the number of ways to form and associate a product of $n + 1$ distinct given factors. Obviously, $A_n = (n + 1)! C_n$. If this doesn't seem to be making any progress, be patient. We have $A_1 = 2$ and we accept $A_0 = 1$. From a product p of $x_1, x_2, x_3, \ldots, x_n$, we get a product of the numbers $x_1, x_2, x_3, \ldots, x_n, x_{n+1}$, either by multiplying p by x_{n+1} on either side of p or else by multiplying either of the 2 factors of 1 of the $n - 1$ multiplications in p on either side of the factor by x_{n+1}. Thus we obtain a total of $2 + 4(n - 1)$, or $4n - 2$, products, and since every product of the $n + 1$ numbers can be obtained in this way, we have the recurrence relation

$$A_n = (4n - 2)A_{n-1} \text{ for } n > 0 \text{ with } A_0 = 1.$$

Have we made progress? Yes. Unwinding this recurrence relation, we get

$$\begin{aligned}
A_n &= (4n - 2)(4n - 6)(4n - 10) \cdots (10)(6)(2) \\
&= 2^n (2n - 1)(2n - 3)(2n - 5) \cdots (5)(3)(1) \\
&= 2^n (2n)! / [(2n)(2n - 2)(2n - 6) \cdots (6)(4)(2)] \\
&= 2^n (2n)! / [2^n n!] \\
&= (2n)! / n!.
\end{aligned}$$

Thus,

$$A_n = \frac{(2n)!}{n!} \text{ for } n \geq 0$$

(and we can see why we accepted A_0 as 1). Further, we then have $C_n = A_n/(n+1)! = (2n)!/[n!n!(n+1)]$. Thus,

$$C_n = \frac{1}{n+1}\binom{2n}{n} \text{ for } n \geq 0.$$

The numbers C_n are called **Catalan numbers**.

n:	0	1	2	3	4	5	6	7	8	9	10
C_n:	1	1	2	5	14	42	132	429	1430	4862	16796

From the formula above, we can easily derive another recurrence relation for computing successive Catalan numbers.

$$C_{n+1} = \frac{2(2n+1)}{n+2}C_n \text{ for } n > 0 \text{ with } C_0 = 1.$$

The Catalan numbers pop up here and there. Given a convex polygon having $n+2$ sides, suppose that there are c_n ways to join the vertices to triangulate the polygon by nonintersecting diagonals. So, $c_1 = 1$, and we take $c_0 = 1$. Let the vertices of the polygon be $V_0, V_1, V_2, \ldots, V_{n+1}$. The side with endpoints V_0 and V_{n+1} belongs to some triangle with third vertex, say V_k such that $1 \leq k \leq n$. The convex polygon with vertices $V_0, V_1, \ldots, V_k$ has $k+1$ sides and can be triangulated in c_{k-1} ways. The convex polygon with vertices $V_k, V_{k+1}, \ldots, V_{n+1}$ has $n-k+2$ sides and can be triangulated in c_{n-k} ways. Thus, there are $c_{k-1}c_{n-k}$ possible triangulations when one of the triangles has vertices V_0, V_k, V_{n+1} Hence, $c_n = \sum_{k=1}^{n} c_{k-1}c_{n-k}$ for $n \geq 1$ with $c_0 = 1$. Since $\{c_n\}$ satisfies a Catalan recurrence relation and its initial condition, then $c_n = C_n$. That is, $c_n = \frac{1}{n-1}\binom{2n}{n}$. Euler found this formula in 1758. The number of ways to draw n nonintersecting cords whose endpoints are $2n$ given points on a circle is also seen to be C_n.

Perhaps the most interesting occurrence of the Catalan numbers is in finding the number of ways to flip a coin $2n$ times to get n heads and n tails with the number of heads flipped at any time not exceeded by the number of tails flipped at that time. (This is, admittedly, a digression from our topic of recurrence relations, to which we will return with Example 20.) We generalize the problem to count the number of ways to toss h heads and t tails with the number of tails tossed never exceeding the number of heads tossed. Possibly we are thinking of a "head" as a score for the Home Team in a game or maybe a vote coming in for the candidate Honest Helen in the election returns.

In any case, our model is a sequence of h H's and t T's. For our immediate purpose, here and below, we will call a sequence of H's and T's "desirable"

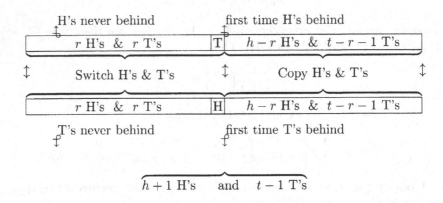

FIGURE 6.3. The Switch and Copy Trick.

if at every point in the sequence the number of H's is not exceeded by the number of T's up to that point. For example, HHHTTT is desirable but HTTHHH is not. It is the desirable sequences that we want to count. We assume $h \geq t > 0$. Of the possible $\binom{h+t}{h}$ sequences having h H's and t T's, we count those for which our condition fails. We can then subtract to obtain the number of desirable sequences. Even these "undesirable" sequences are not easy to count. We change them into something else that is easy to count. Taking an undesirable sequence, where the first time the number of T's exceeds the number of H's is in the $(2r+1)^{\text{st}}$ spot, with $0 \leq r$, we switch all the H's and T's in the first $2r+1$ spots and copy the rest of the sequence. We now have a sequence of $h + 1$ H's and $t - 1$ T's. Of course, this does no good if we cannot go backwards from such a sequence to one of our undesirable sequences. We can; see Figure 6.3. Since $h + 1 > t - 1$, at some point in any sequence of $h + 1$ H's and $t - 1$ T's the number of H's must first exceed the number of T's. Let this be at the $(2r+1)^{\text{st}}$ spot. The switch and copy trick of switching the first $2r+1$ spots and copying the rest brings us to one of our undesirable sequences—that is, a sequence of h H's and t T's such that at some point the number of T's exceeds the number of H's. We have a one-to-one correspondence between the set of our undesirable sequences, which we want to count, and the set of our new sequences, which is the set of all sequences consisting of $h+1$ H's and $t-1$ T's. There are obviously $\binom{h+t}{h+1}$ of these new sequences. Thus, the number of desirable sequences here is the difference between $\binom{h+t}{h}$ and $\binom{h+t}{h+1}$. The most interesting case is when $h = t$.

Observation. The number of sequences of h H's and t T's such that $h \geq t \geq 0$ and such that the number of H's is not exceeded by the number of T's at any point in the sequence is

$$\frac{h-t+1}{h+1} \binom{h+t}{h}.$$

When $h = t = n$, this number is C_n.

In blockwalking from $(0,0)$ to (n,n), with a step in the x-direction equally as likely as a step in the y-direction, what is the probability of never crossing above the line with equation $Y = X$? Our desirable sequences above model the problem with H interpreted as a step in the x-direction and a T as a step in the y-direction. Therefore, there are C_n paths that are never above $Y = X$. We know that there are $\binom{2n}{n}$ blockwalking paths from $(0,0)$ to (n,n). Hence, the desired probability is $\frac{1}{n+1}$.

Suppose $h > t$. Drop the mandatory initial H from a sequence of h H's and t T's with the property that in front of each letter except the first the number of H's exceeds the number of T's. We then have a sequence of $h-1$ H's and t T's such that at each point in the sequence the number of H's is not exceeded by the number of T's. These are our desirable sequences from above. Conversely inserting an H in front of a desirable sequence from above produces a sequence with the property that in front of any letter except the first there are more H's than T's. So, by the Observation above, there are $\frac{(h-1)-t+1}{(h-1)+1}\binom{(h-1)+t}{h-1}$ sequences with the new property. In the case $h = t = n$, we can drop the mandatory initial H and the mandatory terminal T to get one of the desirable sequences from above consisting of $h-1$ H's and $t-1$ T's. Again, we can reverse the process. Here, the count of sequences with the new property is $\frac{(n-1)-(n-1)+1}{(n-1)+1}\binom{(n-1)+(n-1)}{n-1}$.

Corollary. If $h > t$, then the number of sequences of h H's and t T's such that in front of each letter except the first the number of H's exceeds the number of T's is

$$\frac{h-t}{h} \binom{h+t-1}{t}.$$

If $h = t = n$, then the number of sequences of n H's and n T's such that in front of each letter except the first the number of H's exceeds the number of T's is C_{n-1}.

We now want to count the set of sequences of h H's and t T's that are attached to the end of a sequence of q H's such that at every point in the total sequence of length $h + q + t$ the number of H's is never exceeded by the number of T's. We are looking at the desirable sequences of $h + q$ H's and t T's that begin with at least q H's. The number of all sequences of

$h + q$ H's and t T's that begin with at least q H's is $\binom{h+t}{h}$. We suppose that $q \geq 0$. We assume that $h + q \geq t$, as otherwise the count is 0, and also assume that $t > q$, as otherwise the count is clearly $\binom{h+t}{h}$. This time an undesirable sequence has the form

| q H's | $r - q$ H's & r T's | T | $h - r + q$ H's & $t - r - 1$ T's |

with the lone "T" being the first occurrence where the number of H's trails the number of T's. Use a trick often enough and it becomes a method. The switch and copy trick gives a sequence of the form

| q T's | r H's & $r - q$ T's | H | $h - r + q$ H's & $t - r - 1$ T's |

consisting of $h + q + 1$ H's and $t - 1$ T's and beginning with at least q T's. Conversely, each such sequence is the result of a switch and copy trick applied to an undesirable sequence that begins with at least q H's. The number of undesirable sequences this time is the same as the number of all sequences of $h + q + 1$ H's and $t - 1 - q$ T's.

Observation. With $h + q \geq t > q \geq 0$, the number of sequences of length $h + q + t$ consisting of h H's and t T's attached to the end of a sequence of q H's with the number of H's at every point never exceeded by the number of T's up to that point is

$$\binom{h+t}{h} - \binom{h+t}{h+q+1}.$$

Example 20. How many ternary sequences of length n have no 0 and 1 adjacent?

We let a_n be the number of the desired sequences of length n. We are in trouble from the beginning, when we consider counting the sequences that begin with a 0. The next digit could be a 0 or a 2, but then we are back to considering those that begin with 00. Now we are going in circles. The trick here is to let z_n denote the number of such sequences of length n that begin with a 0, which is also the number of such sequences of length n that begin with a 1. This last observation is true by the symmetry of 0 and 1 in the problem. In other words, we could interchange all the 0's with all the 1's and accomplish nothing. We now have the following figures to consider.

0........ z_n	00........ z_{n-1}
1........ z_n	01........ 0
2........ a_{n-1}	02........ a_{n-2}
So $a_n = 2z_n + a_{n-1}$.	So $z_n = z_{n-1} + a_{n-2}$.

We have 2 recurrence relations but we want to eliminate the z's. We rewrite the first as $2z_n = a_n - a_{n-1}$ and substitute this relation in our second

$2z_n = 2z_{n-1} + 2a_{n-2}$, which was multiplied by 2 only to avoid fractions, to get

$$[a_n - a_{n-1}] = [a_{n-1} - a_{n-2}] + 2a_{n-2}.$$

This last equation simplifies to

$$a_n = 2a_{n-1} + a_{n-2}.$$

We calculate the initial conditions $a_0 = 1$ and $a_1 = 3$ to finish the problem. We can use $a_2 = 3^2 - 2$ as a check.

Example 21. How many ternary sequences of length n have neither consecutive 0's nor consecutive 1's?

We let a_n be the number of ternary sequences of length n that have neither consecutive 0's nor consecutive 1's. We let b_n, c_n, d_n, be the number of such sequences that end in 0, 1, 2, respectively. If we consider the last 2 terms of such sequences, the figure

$\ldots 00$	0	$\ldots 10$	c_{n-1}	$\ldots 20$	d_{n-1}
$\ldots 01$	b_{n-1}	$\ldots 11$	0	$\ldots 21$	d_{n-1}
$\ldots 02$	b_{n-1}	$\ldots 12$	c_{n-1}	$\ldots 22$	d_{n-1}

together with the observation that $d_n = a_{n-1}$ leads us to

$$\begin{aligned}
a_n &= 2b_{n-1} + 2c_{n-1} + 3d_{n-1} \\
&= 2[b_{n-1} + c_{n-1}] + 3d_{n-1} \\
&= 2[a_{n-1} - d_{n-1}] + 3d_{n-1} \\
&= 2a_{n-1} + d_{n-1} \\
&= 2a_{n-1} + a_{n-2}.
\end{aligned}$$

We calculate the initial conditions $a_0 = 1$ and $a_1 = 3$ to finish the problem. We can use $a_2 = 3^2 - 2$ as a check.

You may want to explain why Example 20 and Example 21 have the same solution.

Modeling Problems.

Model each of the following problems with a recurrence relation and its initial conditions.

1. Let a_n be the number of ternary sequences of length n that have no consecutive digits equal.

2. Let a_n be the number of ternary sequences of length n that have an even number of 0's.

3. There are a_n ways of selecting n 2-person committees from $2n$ persons.

4. There are a_n ways of selecting n 4-person committees from $4n$ persons.

5. Let a_n be the number of ways to stack n poker chips, each of which can be red, white, blue, or green, such that there is a green chip immediately on top of each red chip.

6. An experiment consists of flipping a coin until the second head appears. Let a_n be the number of experiments that require at most n flips.

7. Let a_n be the number of quaternary sequences of length n that have an even number of 0's.

8. How many ternary sequences of length n have no 1 immediately to the right of any 0?

9. How many ternary sequences of length n have no double zero?

10. How many ternary sequences of length n have no 1 anywhere to the right of any 0?

11. How many ternary sequences of length n have no 1 appearing immediately to the right of some 0?

12. How many quaternary sequences of length n have a 1 and are such that the first 1 precedes any 0?

13. How many ways can a 2-by-n rectangular board be tiled with 1 by 2 and 2-by-2 rectangular pieces?

14. How many ways are there to stack n poker chips, each of which can be red, white, blue, or green, under the following condition?

    (a) No green chip is (directly) on top of any red chip.

    (b) No green chip is (directly) on top of some red chip.

    (c) A green chip is always (somewhere) above some red chip.

    (d) A green chip is always (somewhere) above any red chip.

    (e) A green chip is never (anywhere) above some red chip.

15. How many ways are there to arrange $2n$ persons of different heights into an array of 2 rows and n columns such that heights increase in each row and in each column?

Practice Exam.

You have 2 hours. You may omit or miss 3 questions without penalty. For each of the following 10 conditions, give a recurrence relation with initial conditions that models the solution to the question, How many ways are there to form an n-digit quaternary sequence that satisfies the given condition?

1. No 1 precedes any 0.

2. Any 1 is followed immediately by a 0.

3. There is a subsequence 00.

4. There are no adjacent 0's and no adjacent 1's.

5. No 1 is adjacent to any 0.

6. Any 0 is adjacent to another 0.

7. Any 1 is adjacent to at least one 0.

8. There is no subsequence 000.

9. There is no subsequence 001.

10. There is no subsequence 010.

We end the section with a mystery. Assume g is the generating function for $\{C_n\}$. So, $g = \sum_{r=0}^{\infty} C_r z^r$. Now, since $C_{n+1} = \sum_{i=0}^{n} C_i C_{n-i}$ for $n > 0$ with $C_0 = 1$, we have

$$g - C_0 = \sum_{r=1}^{\infty} C_r z^r = \sum_{n=0}^{\infty} C_{n+1} z^{n+1} = z \sum_{n=0}^{\infty} \left[\sum_{i=0}^{n} C_i C_{n-i} \right] z^n = zg^2.$$

So, $g - 1 = zg^2$, or $zg^2 - g + 1 = 0$. Solving this quadratic equation in g, we get

$$g = \frac{1 - \sqrt{1 - 4z}}{2z}.$$

Everything is straightforward up to the last step. Can a quadratic in g be solved in this way? Does $\sqrt{1 - 4z}$ even make sense? What does it all mean?

§47. Nonhomogeneous Recurrence Relations

Occasionally one needs to solve a recurrence relation of the form

$$a_n = c_1 a_{n-1} + c_2 a_{n-2} + c_3 a_{n-3} + \cdots + c_p a_{n-p} + g(n)$$

for $n \geq p$, with given constants c_i, where $g(n)$ is a given function of n. If $g(n)$ is zero, the recurrence relation is called **homogeneous**, and, otherwise, **nonhomogeneous**. Our approach here will be a cookbook approach and be explained mostly by doing an example. We will call the recurrence relation $a_n = c_1 a_{n-1} + c_2 a_{n-2} + c_3 a_{n-3} + \cdots + c_p a_{n-p}$ "the homogeneous part" of the nonhomogeneous recurrence relation above.

The principal observation is that the difference of 2 solutions to the general nonhomogeneous recurrence relation is a solution to the homogeneous part of the recurrence relation. The application of this is that if we can find the general solution to the homogeneous part and a particular solution to the nonhomogeneous recurrence relation, then we can combine these to get a general solution to the nonhomogeneous recurrence relation.

As our illustrative example, we take

$$a_n = 2a_{n-1} + 3a_{n-2} - 8n^2 + 9 \cdot 2^{n-1}$$

for $n \geq 1$ with $a_0 = a_1 = 1$.

We first find a general solution to the homogeneous part. In our example, we want the general solution to

$$a_n = 2a_{n-1} + 3a_{n-2}.$$

Since the roots of the characteristic equation are -1 and 3, we have the general solution

$$a_n = b(-1)^n + c(3)^n$$

to the homogeneous part, with b and c arbitrary constants.

To make the algebra a little easier, we separate the polynomial part and the exponential part of $g(n)$ in our example. We turn to finding a particular solution of

$$a_n = 2a_{n-1} + 3a_{n-2} - 8n^2.$$

If $g(n)$ is a polynomial of degree m, then we should try a polynomial of degree m for a particular solution; although, if the recurrence relation is of the form $a_n = a_{n-1} + g(n)$, a polynomial of degree $n+1$ will be necessary. In our case, $-8n^2$ is a polynomial of degree 2 and so we try for a particular solution of the form $dn^2 + en + f$. We need to determine the constants d, e, f such that

$$[dn^2 + en + f]$$
$$= 2[d(n-1)^2 + e(n-1) + f] + 3[d(n-2)^2 + e(n-2) + f] - 8n^2.$$

This reduces to the equation

$$0 = (4d - 8)n^2 + (4e - 16d)n + (14d - 8e + 4f).$$

Since this last equation is to hold for all values of n, then the 3 coefficients must each be 0. So, we get $d = 2$, $e = 8$, and $f = 9$. Thus,

$$a_n = 2n^2 + 8n + 9$$

is a particular solution to $a_n = 2a_{n-1} + 3a_{n-2} - 8n^2$.

We next turn to finding a particular solution to

$$a_n = 2a_{n-1} + 3a_{n-2} + 9 \cdot 2^{n-1}.$$

If $g(n)$ is a multiple of t^n for some constant t, then we try $k \cdot t^n$, with k a constant that determines a particular solution. This usually works, but, if t is already a root with multiplicity r of the characteristic equation of the homogeneous part, then we try $kn^r t^n$. In our example, $9 \cdot 2^{n-1}$ is a multiple of 2^n and we try $a_n = k2^n$ as a particular solution. So, take

$$[k2^n] = 2[k2^{n-1}] + 3[k2^{n-2}] + 9 \cdot 2^{n-1}$$

to get $k = -6$. Hence,

$$a_n = -3 \cdot 2^{n+1}$$

is a particular solution to $a_n = 2a_{n-1} + 3a_{n-2} + 9 \cdot 2^{n-1}$.

Putting all this together, we see that

$$a_n = b(-1)^n + c3^n + 2n^2 + 8n + 9 - 3 \cdot 2^{n+1}$$

is a general solution to our original nonhomogeneous recurrence relation. Finally, we use the given initial conditions $a_0 = a_1 = 1$ to determine that $b = 0$ and $c = -2$. Therefore, our final solution is the equation

$$a_n = 2n^2 + 8n + 9 - 3 \cdot 2^{n+1} - 2 \cdot 3^n \text{ for } n \geq 0.$$

We finish with 1 more example, which will provide a useful formula. Consider the recurrence relation $a_n = a_{n-1} + n^2$ for $n > 0$ with $a_0 = 0$. Unwinding this when $n > 0$ shows that a_n is the sum of the first n squares. In other words, $a_n = \sum_{k=0}^{n} k^2$ for $n \geq 0$. Applying the suggested techniques above, we expect that this recurrence relation has a general solution that is a cubic in n. So we try $a_n = bn^3 + cn^2 + dn + e$ to get

$$[bn^3 + cn^2 + dn + e] = [b(n-1)^3 + c(n-1)^2 + d(n-1) + e] + n^2,$$

which, in turn, gives us

$$0 = (1 - 3b)n^2 + (3b - 2c)n + (c - b - d).$$

So, $b = 1/3$, $c = 1/2$, and $d = 1/6$, with e arbitrary. Thus, we have the general solution $a_n = (2n^3 + 3n^2 + n)/6 + e$. The initial condition $a_0 = 0$

implies $e = 0$. We have solved our recurrence relation and have the following formula.

Observation.

$$\sum_{k=1}^{n} k^2 = \frac{n(n+1)(2n+1)}{6}.$$

Homework.

1. Give a closed formula for each of the following recurrence relations with their initial conditions.

   (a) Solve $a_n = a_{n-1} + 6a_{n-2} + 3n$ when $a_0 = a_1 = 0$.

   (b) Solve $a_n = a_{n-1} + 6a_{n-2} + 2^n$ when $a_0 = a_1 = 0$.

   (c) Solve $a_n = a_{n-1} + 2a_{n-2} + 2n$ when $a_0 = a_1 = 0$.

   (d) Solve $a_n = a_{n-1} + 2a_{n-2} + 2^n$ when $a_0 = a_1 = 0$.

   (e) Solve $a_n = 4a_{n-1} - 4a_{n-2} + 2^n$ when $a_0 = a_1 = 1$.

2. Show that the total number of triangles in an equilateral triangle of side n tiled by equilateral triangles of side 1 is given by the formula

$$\frac{4n^3 + 10n^2 + 4n - 1 + (-1)^n}{16}.$$

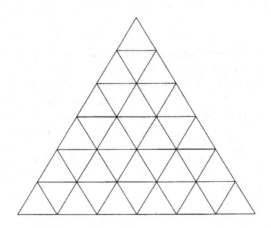

7
Mathematical Induction

§48.   The Principle of Mathematical Induction

A very powerful method of proof for a statement involving integers is analogous to climbing a ladder. If we (1) can get on the first rung of the ladder and (2) if we can get from each rung to the next higher rung, then we claim that we can get to every rung of the ladder. This may seem like common sense. It is, and this common sense is codified as follows, where by a "proposition" we mean any statement that makes sense.

The Principle of Mathematical Induction. For each positive integer n, suppose that $P(n)$ is a proposition. If

- $P(1)$ is true and

- $P(k)$ is true implies $P(k+1)$ is true, for each $k \geq 1$,

then $P(n)$ is true for each positive integer n,

We do not prove the Principle of Mathematical Induction. It is essentially an axiom of mathematics. Specifically, it is an immediate consequence of Peano's axioms for the definition positive integers. The application of the Principle of Mathematical Induction is called *mathematical induction* and is a 3-step process: Given statement $P(n)$, the first step is to prove that $P(1)$ is true. This is often called the *basis step*. The second step is to prove that $P(k)$ implies $P(k+1)$ whenever $k \geq 1$. The second step is called the *induction step*, and $P(k)$ in this step is called the *induction hypothesis*.

The third step is the appeal to the Principle of Mathematical Induction in order to conclude that $P(n)$ is true for all positive integers n. Since there is "nothing to compute" for the third step, this essential step is too frequently overlooked. We will not follow those who ignore this last step because they think that it is so obvious that it need not be mentioned.

The induction step can also be stated as "$P(k-1)$ is true implies $P(k)$ is true, for each $k > 1$." This is a change in notation only, and is not a change in content.

The name of mathematical induction is unfortunate because the name is misleading. Since induction is a process of forming generalizations extrapolated from some examples, we avoid the jargon that uses "induction" in place of "mathematical induction," based on the premise that "everybody knows" the intended meaning. Mathematical induction has little to do with induction. We might use induction, as in our first example, to guess a proposition that we then prove by mathematical induction. Such guesses are propositions and, when we feel they are correct, we call them conjectures. Our first example arises from the pattern that results from summing the first n odd integers. We observe that

$$1 = 1^2,$$
$$1 + 3 = 2^2,$$
$$1 + 3 + 5 = 3^2.$$
$$1 + 3 + 5 + 7 = 4^2,$$
$$1 + 3 + 5 + 7 + 9 = 5^2,$$

and by induction guess that $\sum_{j=1}^{n}(2j-1) = n^2$ for each positive integer n. We see a pattern and conjecture that the pattern holds for all the positive integers. Our proposition may be true or our proposition may be false. We prove that our conjecture is actually true by mathematical induction. Our proposition $P(n)$ is the statement: $\sum_{j=1}^{n}(2j-1) = n^2$. First we must prove the basis step that $P(1)$ is true. This is easy enough, since $2 - 1 = 1^2$. For the induction step, we assume that $\sum_{j=1}^{k}(2j-1) = k^2$ for some arbitrary integer k such that $k \geq 1$. In other words, we have assumed that $P(k)$ is true and we must now argue that $P(k+1)$ is true. The calculation

$$\sum_{j=1}^{k+1}(2j-1) = \left[\sum_{j=1}^{k}(2j-1) \right] + [2(k+1)-1] = k^2 + (2k+1) = (k+1)^2$$

proves the induction step. Note that proving the induction step is actually a *deduction*. From the assumption

$$1 + 3 + 5 + \cdots + (2k-1) = k^2$$

for some $k \geq 1$ and the incontestable equation

$$2(k+1) - 1 = (k+1)^2 - k^2,$$

we deduce that

$$1 + 3 + 5 + \cdots + [2k - 1] + [2(k + 1) - 1] = (k + 1)^2.$$

We therefore conclude by the Principle of Mathematical Induction that $P(n)$ is true for each positive integer n. Thus, we have proved

$$\sum_{j=1}^{n}(2j - 1) = n^2$$

for each positive integer n. A direct proof of this formula follows from knowing the formula for the sum of the first n positive integers. For each n, with $n \geq 1$,

$$\sum_{j=1}^{n}(2j - 1) = 2\sum_{j=1}^{n}j - \sum_{j=1}^{n}1 = 2\frac{n(n + 1)}{2} - n = n^2.$$

We give 3 examples illustrating that induction by itself is not a proof at all. Given the formula $f(n) = n^2 + n + 41$, if we compute $f(1)$, $f(2)$, $f(3)$, $f(4)$, $f(5)$, ... , $f(39)$, we always get a prime number. It is certainly reasonable to assume that this pattern will continue. Reasonable, yes; but also incorrect. In fact, $f(40) = 41^2$ and 41^2 is obviously not a prime. Our favorite example is next. From the equations

$$x^1 - 1 = x - 1,$$
$$x^2 - 1 = (x - 1)(x + 1),$$
$$x^3 - 1 = (x - 1)(x^2 + x + 1),$$
$$x^4 - 1 = (x - 1)(x + 1)(x^2 + 1),$$
$$x^5 - 1 = (x - 1)(x^4 + x^3 + x^2 + x + 1),$$
$$x^6 - 1 = (x - 1)(x + 1)(x^2 + x + 1)(x^2 - x + 1),$$
$$x^7 - 1 = (x - 1)(x^6 + x^5 + x^4 + x^3 + x^2 + x + 1),$$
$$x^8 - 1 = (x - 1)(x + 1)(x^2 + 1)(x^4 + 1),$$
$$x^9 - 1 = (x - 1)(x^2 + x + 1)(x^6 + x^3 + 1),$$
$$x^{10} - 1 = (x - 1)(x + 1)(x^4 + x^3 + x^2 + x + 1)(x^4 - x^3 + x^2 - x + 1),$$

$$\vdots$$

it looks like all the coefficients of the real factors of $x^n - 1$ are ± 1. This is indeed the case for each of the first 104 positive integers. However, a factor of $x^{105} - 1$ has a -2 as a coefficient. The third and last example is even more extraordinary. If we start calculating the values of $991n^2 + 1$ for the positive integers, we seem not to ever get a square number. We can gather a lot of convincing evidence that this will always be the case. A lot. Actually, this will be true until our number n reaches

$$12,055,735,790,331,359,447,442,538,767$$

when we do get a square. Even very much larger examples are available, but we get the idea. Induction does usually work but not always. It is the uncertainty that requires us to produce deductive proofs for our conjectures.

As an easy illustration, let's use mathematical induction to prove the proposition $P(n)$ that states that the sum of the first n positive integers is $n(n+1)/2$. Certainly $1 = 1(2)/2$. So $P(1)$ is true. We then show that $P(k)$ implies $P(k+1)$ for $k \geq 1$ by the calculation

$$\sum_{j=1}^{k+1} j = \left[\sum_{j=1}^{k} j \right] + [k+1] = \frac{k(k+1)}{2} + [k+1] = \frac{(k+1)(k+2)}{2}.$$

Hence, by mathematical induction, the formula holds for all n.

Now let's "prove" that the sum of the first n positive integers is also $(n-1)(n+2)/2$. This time we have

$$\sum_{j=1}^{k+1} j = \left[\sum_{j=1}^{k} j \right] + [k+1] = \frac{(k-1)(k+2)}{2} + [k+1]$$

$$= \frac{k(k+3)}{2} = \frac{[(k+1)-1][(k+1)+2]}{2}.$$

So, $P(k)$ does imply $P(k+1)$. What is going on here? You may have noticed that we did not prove the basis step. In this case it is impossible to prove $P(1)$, because $P(1)$ is false. Without the basis step, the induction step is useless. Needless to say, our proposition is itself false.

Here is another flimflam argument. We will "prove" that all balls are the same color- and thus simplify most of the problems we have encountered in this text. Certainly, the balls in any set consisting of exactly 1 ball are all the same color. We assume that the balls in any set of k balls are all the same color. Now, suppose a set of $k+1$ balls is produced. We argue that we can line them up and the first k must be all the same color by our induction hypothesis; the last k must also be all the same color by our induction hypothesis. It follows that all the $k+1$ balls must be of the same color. We claim (with our fingers crossed) that we have proved that all balls are the same color by mathematical induction.

Proving the basis step in a proof by mathematical induction is frequently trivial. This is not always the case, however. For example given $n+1$ arbitrary squares, we want to prove it is possible to cut the squares into

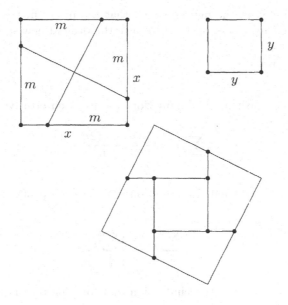

FIGURE 7.1. Dissection of 2 Squares to Form 1 Square.

pieces that can be arranged to form a new square. (For the case $n = 0$, there is nothing to prove. As Gertrude Stein probably never said, A square is a square is a square.) The induction step in a mathematical induction argument is to show that $P(k)$ is true implies $P(k + 1)$ is true for each $k \geq 1$. The argument that if we are given $k + 1$ squares we can take the first k to form a new square by the induction hypothesis and then combine this square with the kth square to form a new square is valid as long as we can take 2 squares and combine them into 1 square. In other words, it is the case of 2 squares, when $n = 1$, that requires some nontrivial proof. The rest follows rather easily by mathematical induction. (It is the case of 2 balls that was not proved in the flimflam argument above. If the balls in every set of 2 balls actually have the same color, then it would, in fact, correctly follow that all balls have the same color.) Here, the case for $n = 1$ is true as can be gleaned from Figure 7.1, where we have 2 given squares of sides x and y with $x \geq y$. We let $m = (x + y)/2$. So $m = (x - m) + y$. We mark off segments of length m on the sides of the larger (or congruent) square as in the figure. The joins of these points on opposite sides of this square are necessarily perpendicular and dissect the square into 4 congruent quadrilaterals (or triangles if $x = y$). (The rotation of 90 degrees about the intersection of these joins permutes the segments of length m and so permutes the quadrilaterals.) After translating each of the quadrilaterals to attach them to the sides of the smaller square in the manner indicated in the figure, we see that the pieces fit together (this is where we need $m = (x - m) + y$) to form the desired new square.

Homework MI I.

1. Prove by mathematical induction that for each positive integer n

$$\sum_{j=1}^{n} j^2 = \frac{n(n+1)(2n+1)}{6}.$$

2. Prove by mathematical induction that for each positive integer n

$$\sum_{j=1}^{n} j^3 = \left[\sum_{j=1}^{n} j \right]^2.$$

3. Prove by mathematical induction that for each positive integer n

$$\sum_{j=1}^{n} (2j-1)^2 = \frac{n(2n-1)(2n+1)}{3}.$$

4. Prove by mathematical induction that for each positive integer n

$$\sum_{j=1}^{n} (2j-1)^3 = n^2(2n^2 - 1).$$

5. Guess and prove a formula for $\sum_{j=1}^{n} j! j$ for positive integer n.

6. Prove that when n dice are rolled the number of possible outcomes having an even sum equals the number of possible outcomes having an odd sum. (This is a tricky question.)

§49.   The Strong Form of Mathematical Induction

We next use mathematical induction to prove the validity of the following argument, which is called the *strong form of mathematical induction* because the induction hypothesis here has stronger requirements.

Strong Form of Mathematical Induction. For each positive integer n, suppose that P(n) is a proposition. If

- P(1) is true and

- P(1), P(2), ... , P(k) are true implies P($k+1$) is true, for each $k \geq 1$,

then P(n) is true for each positive integer n.

To prove the strong form of mathematical induction, we let Q(n) be the proposition, "P(m) is true for every positive integer m such that $1 \leq m \leq n$." Let's be clear what we have to do. We suppose [1] the ordinary Principle of Mathematical Induction, [2] P(1) is true, and [3] P(1), P(2), ... , P(k) are true implies P($k+1$) is true, for each $k \geq 1$. We must show that Q(n) is true for each positive integer n. Now, Q(1) is true because P(1) is true by [2]. We assume Q(k) is true for some arbitrary k such that $k \geq 1$. Hence, P(m) is true for m such that $1 \leq m \leq k$. Therefore, P(1), P(2), ... , P(k), and P($k+1$) are true by [3]. That is, Q($k+1$) is true. This completes the induction step showing that Q(k) implies Q($k+1$) for $k \geq 1$. Therefore, by [1], Q(n) is true for all positive integers n. Hence, P(n) is true for all positive integers n, as desired.

In the next section we will use the strong form to prove the principal theorem of the chapter. The usual example that is used to show the strength of the strong form of mathematical induction over ordinary mathematical induction is the proof that each integer greater than 1 can be factored into a product of primes. First, note that ordinary induction here is of little use since knowing how to factor k does not usually help at all in trying to factor $k + 1$. (For example, take $k = 2^{64}$.) To fit the form above exactly, we need to state our proposition P(n) as "$n + 1$ is a product of primes for positive integer n." Anyway, 2 is a prime and so is a product of primes. (In this case the product has only 1 factor.) So, P(1) is true. For the induction step, we assume P(1), P(2), ... , and P(k) are true for some k such that $k \geq 1$. Thus, if $2 \leq r \leq k + 1$, then r is a product of primes. We need to show that $k + 2$ is a product of primes. If $k + 2$ is itself a prime, then we are done. Suppose $k + 2$ is not a prime. Then, there are integers s and t such that $k + 2 = st$ with $2 \leq s < k + 2$ and $2 \leq t < k + 2$. Since both s and t are products of primes by the induction hypothesis, then their product is also a product of primes. Thus, by the strong form of mathematical induction, we conclude that every positive integer greater than 1 is a product of primes. (If we go on to show that the primes are necessarily unique, up to order of the factors, the result is called the Fundamental Theorem of Arithmetic.)

The example above already touches on an adjustment that can be made to mathematical induction. We really want to start counting with 2 in the example above. If we are talking about properties of polygons, we usually want to start with the case $n = 3$ for obvious reasons. It is also evident that we very frequently want to start with the case $n = 0$. Thus, without proof, we state that the following are valid arguments for proving propositions concerning integers. The first merely shifts what is called the *base* of the mathematical induction from 1 to b and is the special case $r = 1$ of the second. In each case, we still call the first requirement the basis step and the second the induction step.

Mathematical Induction. Suppose b is an integer. Suppose, for each integer n such that $n \geq b$, that $P(n)$ is a proposition. If

- $P(b)$ is true and

- $P(k)$ is true implies $P(k+1)$ is true, for each $k \geq b$,

then $P(n)$ is true for each integer n such that $n \geq b$.

Complete Mathematical Induction. Suppose that b is an integer, that r is a positive integer, and, for each integer n such that $n \geq b$, that $P(n)$ is a proposition. If

- $P(b)$, $P(b+1)$, $\ldots$, $P(b+(r-1))$ are true and

- $P(k-r)$, $\ldots$, $P(k-2)$, $P(k-1)$ are true implies $P(k)$ is true whenever $k - r \geq b$,

then $P(n)$ is true for each integer n such that $n \geq b$.

There are many forms of mathematical induction. The form that we have called Complete Mathematical Induction is especially applicable to recurrence relations. As an example, we prove that the recurrence relation

$$a_0 = a_1 = 1 \quad \text{and} \quad a_n = a_{n-1} + 2a_{n-2} \text{ for } n \geq 2$$

has the solution

$$a_n = \frac{2^{n+1} + (-1)^n}{3} \text{ for } n \geq 0.$$

We use Complete Mathematical Induction with the base b where $b = 0$. Here, $r = 2$. The basis step requires showing that a_0 and a_1 are correctly given by the formula. This is easily verified:

$$a_0 = 1 = \frac{2+1}{3} \quad \text{and} \quad a_1 = 1 = \frac{2^2 - 1}{3}.$$

For the induction step, we must show that $P(k-2)$ is true and $P(k-1)$ is true implies $P(k)$ is true for $k \geq 2$. That is, our formula satisfies the equation

$$a_k = a_{k-1} + 2a_{k-2} \text{ for } k \geq 2.$$

This is an easy algebraic calculation; all we have to do is check that

$$\frac{2^{k+1} + (-1)^k}{3} = \frac{2^k + (-1)^{k-1}}{3} + 2\frac{2^{k-1} + (-1)^{k-2}}{3}$$

for $k \geq 2$. Therefore, the formula for the solution is valid for all nonnegative integers n.

Homework MI II.

1. Prove that n concurrent planes, with no line on any 3 of the planes, divide space into $n(n-1)+2$ regions, for each positive integer n. (That the planes are concurrent means that they pass through a common point. Note that the sequence 2^1, 2^2, 2^3 gets things off to a roaring start but, alas, then drops dead. The fourth plane does not intersect all of the regions created by the first 3 planes, except at the point of concurrency. Playing with some more values of n should produce a picture that motivates the induction step of a proof by mathematical induction, if not the formula itself.)

2. Show that postage stamps of value 3 cents and 5 cents are sufficient to post any letter requiring more that 7 cents in postage.

3. Prove DeMoivre's Theorem: If n is a nonnegative integer, then

$$(\cos \alpha + i \sin \alpha)^n = \cos n\alpha + i \sin n\alpha.$$

4. If n is an integer such that $n > 4$, then

$$n^2 < 2^n < n!.$$

5. Let b is any positive integer greater than 1. Prove that every positive integer can be uniquely represented in the form

$$r_0 b^0 + r_1 b^1 + r_2 b^2 + \cdots + r_{n-1} b^{n-1} + r_n b^n,$$

where $r_n \neq 0$ and $0 \leq r_j < b$ for $j = 1, 2, \ldots, n$. (Assume—or, if you insist, prove by mathematical induction—the Division Algorithm: If a and b are positive integers, then there exist an integer q (the quotient) and an integer r (the remainder) such that $a = qb + r$ and $0 \leq r < b$. The quotient and remainder are unique.)

6. Prove that every positive integer can be uniquely expressed as a sum of the form

$$c_1 1! + c_2 2! + c_3 3! + \cdots + c_{n-1}(n-1)! + c_n n!$$

where c_j is an integer such that $0 \leq c_j \leq j$ for $j = 1, 2, \ldots, n$.

§50. Hall's Marriage Theorem

We suppose that we have been cast in the roll of matchmaker. We are to match each of the 7 boys in column 1 of Table 7.1, with 1 of the girls he knows, given that the girls known to boy i are listed in the set in the ith row of column 2 in the table. For example, boy 2 knows the girls d, e, and f. In this case we see that our task is impossible. The 4 boys 2, 4, 5, and 6 collectively know only the 3 girls d, e, and f. We are not allowed to match a girl with more than 1 boy. Indeed, if we are to have any hope of success in matching, it is certainly necessary that any collection of k of the boys must collectively know at least k of the girls. What is surprising is that this necessary condition is also a sufficient condition. We will refer to the condition that each collection of k boys must collectively know at least k girls as the *marriage condition*.

We will use the strong form of mathematical induction to prove the sufficiency of the marriage condition. This time our induction step will have the form

$$\text{P}(1), \text{P}(2), \ldots, \text{P}(m-1) \text{ are true implies } \text{P}(m) \text{ is true, for } m > 1,$$

where $\text{P}(k)$ denotes the proposition that there is a matching for k boys iff the marriage condition for k boys holds. Before giving the proof, we mention that our subject is not totally frivolous. We could talk about the *personnel assignment problem*, where we match applicants with positions for which they are variously qualified, or vice versa. One of our exercises will suggest applications to transportation networks.

Hall's Marriage Theorem. Each of n given boys can be matched with a different girl he knows iff, for each k such that for $1 \leq k \leq n$, each collection of k boys collectively know at least k girls.

Proof of Hall's Marriage Theorem. We have mentioned that the necessity of the marriage condition is obvious. If there are some particular k boys that collectively know less than k girls, then a matching of these k boys, in

Boys	Column 2	Column 3	Column 4	Column 5
1	$\{a, b, c, d\}$	$\{1, 2, \mathbf{3}\}$	$\{d, e, f, g\}$	$\{ⓐ, b, c\}$
2	$\{d, e, f\}$	$\{\mathbf{2}, 5\}$	$\{d, e, h\}$	$\{ⓑ, d\}$
3	$\{b, d, g\}$	$\{2, \mathbf{5}\}$	$\{d, g, h\}$	$\{ⓒ, d\}$
4	$\{d, e\}$	$\{2, 3, \mathbf{4}\}$	$\{a, b, c, f\}$	$\{b, ⓓ, e\}$
5	$\{d, f\}$	$\{1, 2, 5\}$	$\{b, c, g, h\}$	$\{b, ⓔ, f, g\}$
6	$\{d, e, f\}$	$\{6, \mathbf{7}\}$		$\{b, c\}$
7	$\{a, b, c, d, g\}$			$\{g\}$

TABLE 7.1. Boys and Girls That Boys Know.

particular, is impossible and, hence, a matching for all the boys is impossible.

The proof of the sufficiency of the condition is by the strong form of mathematical induction on n, the number of boys. For the basis step ($n = 1$), if there is only 1 boy who knows at least 1 girl then a matching is clearly possible. For the induction step, we assume that the proposition is true for $n = 1, 2, \ldots, m - 1$ and then prove that the proposition is true for $n = m$. Thus, we suppose that each set of k boys collectively know at least k girls for $1 \leq k \leq m$. Further (and this is our induction hypothesis), if $1 \leq k < m$, then there is a desired matching of the k boys. We consider 2 cases.

1. Suppose every k boys, with $1 \leq k < m$, collectively know at least $k + 1$ girls (which is more than required by the marriage condition). In this case, we first match some arbitrary boy with some girl he knows. Then we are left with $m - 1$ boys, each k of which collectively know at least k of the remaining girls. There is a matching of these remaining $m - 1$ boys by the induction hypothesis. So each of the m boys is matched with a different girl he knows.

2. Suppose the case above does not hold. Then, there is some k with $1 \leq k < m$ such that there is a set of k boys who collectively know exactly k girls. These k boys can be matched with these k girls by the induction hypothesis. We have left $m-k$ boys to be matched. Consider any set of h of these $m - k$ boys, where $1 \leq h \leq m - k < m$. Now, these h boys must collectively know at least h of the remaining girls, since, otherwise, the h boys together with the k already matched boys would have collectively known fewer than $h + k$ girls, contradicting our assumption. Thus, the remaining $m - k$ boys can be matched by the induction hypothesis. Therefore, each of the m boys can be matched with a girl he knows, as desired.

It follows by the strong form of mathematical induction that the proposition is true for each positive integer n. ∎

This elegant proof and, apparently, the marriage metaphor are due to Halmos and Vaughan (1950). The theorem itself is due to P. Hall (1935). We next state Hall's theorem without the marriage metaphor. Interpret the A_i as the set of acquaintances (girls) that boy i knows and x_i as the girl matched with boy i.

Hall's Theorem (set form). Let $A_1, A_2, \ldots, A_n$ be subsets of set X. Then a necessary and sufficient condition that there exist distinct elements $x_1, x_2, \ldots, x_n$ such that $x_i \in A_i$ is that each union of k of the subsets from among the A_i contain at least k elements, for $1 \leq k \leq n$.

A **transversal** to a list $A_1, A_2, \ldots, A_n$ of sets is a set of distinct representatives of the sets, that is, a set of n distinct elements a_i such that

$a_i \in A_i$ for $i = 1, 2, \ldots, n$. Note that the elements a_i are distinct although this may not be the case for the sets A_i, which are distinguished by their subscripts. For example, $\{1, 2, 3, 4, 5, 7\}$ is a transversal for the 6 sets in column 3 of Table 7.1. We have yet another form of Hall's theorem.

Hall's Theorem (transversal form). Finite sets $A_1, A_2, \ldots, A_n$ have a transversal iff

$$\left| \bigcup_{i \in I} A_i \right| \geq |I|$$

for all I such that $I \subseteq \{1, 2, \ldots, n\}$.

Word has just come down to the matchmaker from the general that the daughters of the general must be among those girls that are matched with the given boys (or heads will roll). Suppose that there is a set D, the daughters of the general, of girls that must be included in the matching of n boys. For each subset I of these boys, the number of girls in D not known by any boy in I cannot exceed the number of boys not included in I. After figuring out that this condition is obviously necessary—and this may take some time—we will show that, again surprisingly, this additional condition along with the marriage condition is sufficient for including the daughters of the general in our matchmaking. First, let's consider an example. In column 4 of Table 7.1, we have the sets A_i of girls known to boy i in row i. There are 5 boys. The daughters of the general are girls a, b, c, and d. The A_i have transversal $\{b, c, d, e, f\}$, which we can see by picking these elements in reverse order from the 5 sets. The question is, Can the daughters of the general be included in a matching? Now, the 3 boys 1, 2, and 3 collectively know the 5 girls d, e, f, g and h, necessarily leaving the 2 boys 4 and 5 to be matched with the 3 girls a, b, and c. Here, 3, the number of the daughters of the general not known by the 3 boys, exceeds 2, the number of the other boys. The task is impossible in this example. Since we value our heads, we had better prove to the general exactly when the task is impossible.

The Daughters-of-the-General Theorem. Suppose n boys collectively know m girls, including the daughters of the general. Each of the n boys can be matched with a different girl he knows such that all the daughters of the general are included in the matching iff, for $1 \leq r \leq n$.

  1. each set of r boys collectively know r girls and

  2. for each set of r boys, the number of daughters of the general not known to any of these r boys does not exceed $n - r$.

Proof. Condition 1 is the marriage condition and is necessary for any matching. Condition 2 is necessary if all the daughters of the general are to be

$$A_1, \quad A_2, \quad \cdots \quad A_n \qquad G \setminus D, \quad G \setminus D, \quad \cdots \quad G \setminus D$$
$$\downarrow \quad\; \downarrow \quad\; \cdots \quad\; \downarrow \qquad\quad \downarrow \qquad\quad \downarrow \qquad\; \cdots \qquad\; \downarrow$$
$$a_1, \quad a_2, \quad \cdots \quad a_n \qquad a_{n+1}, \quad a_{n+2}, \quad \cdots \quad a_m$$

$\underbrace{\qquad\qquad\qquad\qquad\qquad}$ $\underbrace{\qquad\qquad\qquad\qquad\qquad}$

n girls matched with the $m - n$ girls matched with the
original n boys $m - n$ invented boys

FIGURE 7.2. Matching with Daughters of the General.

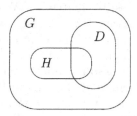

FIGURE 7.3. Daughters of the General.

included in the matching. Our main task is to prove the converse: the 2 conditions together are sufficient.

We introduce the following notation. Let G be the set of all girls. Let D be the set of the daughters of the general. Let A_i be the set of girls known to boy i, for $1 \le i \le n$. So, $G \setminus D$ is the set of all girls who are not daughters of the general. (In general, if S and T are sets, then $S \setminus T$ is the set of all elements in S that are not in T. In particular, if T is a subset of S, then $|S \setminus T| = |S| - |T|$. The symbol "$\setminus$" can be read "setminus.").

We invent $m - n$ additional boys; each of these invented boys knows no daughter of the general but does know all the other girls. We now have m boys and m girls. If we can match all m boys with the m girls, we claim we will have succeeded in matching the original n boys with girls they know such that the daughters of the general are included in this matching. See Figure 7.2. This must be so, since all the m girls are matched but none of the invented boys is matched with a daughter of the general. For our desired matching, all we have to do is ignore the matching of the invented boys.

Suppose we have a set of k of the m boys where $1 \le k \le m$. Our theorem will follow from Hall's Marriage Theorem if we can show that these k boys collectively know at least k girls. Of the k boys, we suppose that h are from the original n boys and that i are from the $m - n$ invented boys. So, $k = h + i$. We may as well assume that $i > 0$, as otherwise we are done by condition 1 in the theorem. Let H be the set of girls collectively known by the h boys. Now, the k boys collectively know the girls in the set $H \cup (G \setminus D)$, which is the set $G \setminus (D \setminus H)$. Figure 7.3 may help. Condition 2 in the statement of the theorem translates as the inequality $|D \setminus H| \le n - h$.

Since $D \setminus H$ is a subset of G, then

$$|H \cup (G \setminus D)| = |G \setminus (D \setminus H)| = m - |D \setminus H| \geq$$
$$m - (n - h) = h + (m - n) \geq h + i = k.$$

Hence, the k boys collectively know at least k girls, as desired. Therefore, there is a matching of the m boys with the m girls. ∎

Our proof of Hall's Marriage Theorem is elegant but does not provide an algorithm for finding a possible matching. We end this section by giving an example that can be generalized to a constructive proof of the theorem. The language is due to Bryant (1993) and fits the marriage metaphor exceedingly well. Refer to column 5 of Table 7.1. In the process of forming a matching for 7 boys, we have so far announced 5 tentative engagements, as denoted by the circled names of the girls in the respective rows. However, when we now come to boy 6 we see that the girls he knows have already been engaged. What are we to do? We throw a party! We invite boy 6 because the party is for him. Boy 6 invites all the girls he knows. These girls invite their fiancés. These boys invite all the girls they know who haven't already been invited. Those girls invite their fiancés. These boys invite all the girls they know who haven't already been invited. ... This process continues until some girl that is not engaged is invited. (The proof of the algorithm hinges on using the marriage condition to argue that there is always such a girl as long as the matching is not complete.) In our example, where we read "→" as "invites," we have

$$\{6\} \to \{b, c\} \to \{2, 3\} \to \{d\} \to \{4\} \to \{e\} \to \{5\} \to \{f, g\}.$$

We stop because girl g, for example, is not engaged. (We ignore that our example is so simple that we see that we should leave girl g for the obligatory match with boy 7 and pick girl f at this point.) Everybody invited to the party comes. Girl g dances with a boy she knows who invited her (boy 5). His fiancée (girl e) is a bit annoyed and so dances with a boy who invited her (boy 4). His fiancée (girl d) is a bit annoyed and so dances with a boy who invited her (boy 2, there is a choice). His fiancée (girl b) is a bit miffed and so dances with a boy who invited her (boy 6). In general, we continue backward through the sets as displayed above until we arrive at the boy who has yet to be matched. In our example, there are 4 couples

$$(6, b), \quad (2, d), \quad (4, e), \quad (5, g)$$

on the dance floor. The dance is so very successful that the dancing couples break off any former engagements and get engaged to their dancing partners. The engagements of the other couples are not affected. So, we have now matched the pairs

$$(1, a), \quad (2, d), \quad (3, c), \quad (4, e), \quad (5, g), \quad (6, b).$$

We have gone from matching the first 5 boys to matching the first 6 boys. Looking at the last row of column 5 in the table, we see that boy 7 knows only girls who are presently engaged. We throw another party, which results in the engagement of each of the 7 boys to a girl he knows. In general, we keep throwing parties until all the boys are engaged.

Homework MI III.

1. If there are n boys and n girls and each collection of k boys collectively know at least k girls, then each collection of k girls collectively know at least k boys.

2. One globber is produced at tremendous cost in each of p distinct origins and 1 globber is in great demand at d distinct destinations. If it is possible to ship a globber from origin x to destination y, then we say that locations x and y are *connected* to each other. Under what circumstances is it possible to ship a globber to each of the d destinations?

3. Deal a shuffled deck of cards into a 4–by–13 array. Now, try to select 1 card from each of the 13 columns to get 1 card of each of the 13 denominations. Is this a good game of solitaire or not?

4. Show that if for a given collection of boys and girls there is a positive integer r such that each boy knows at least r girls and such that each girl knows at most r boys, then there is a matching for the given boys.

5. *The Harem Problem* or "the celebrated problem of the monks." For each of n boys, suppose that boy i wishes to be matched with g_i girls he knows. (Unfairly, each girl can be matched with at most 1 boy.) Find a necessary and sufficient condition for there to be a solution in this case.

6. Prove the following, given n boys and $0 \leq r \leq n$. There exist some r of the n boys that can be matched with different girls they know iff each subset of k of the n boys collectively know at least $k + r - n$ girls. (Hint: Invent $n - r$ additional girls such that each individually knows all the boys.)

7. Suppose that each subset of k from a given set of n boys collectively know at least r girls, with $r > 0$. Show that there are at least $r!$ matchings for the given boys if $r \leq n$. Show that there are at least $\frac{r!}{(r-n)!}$ matchings for the given boys if $r > n$.

8
Graphs

§51. The Vocabulary of Graph Theory

In an undergraduate mathematics curriculum, graphs are traditionally introduced in a course on discrete mathematics, if at all. The theory of graphs is a recent and growing branch of mathematics. The topic is well worth a semester course. However, time is short and our goal here is only to introduce the elementary terminology so that we will have some idea of what "graph theory" means. Of course the first question has to be, What is a graph? As we will see, that is not an easy question to answer; not because the question is so difficult, but because the terminology is not only not standard, it is all over the place. It is easy to say a couple things about what we are not going to talk about. We are not talking about bar graphs or pie charts. We are also not talking about the graphs of functions. For example, parabolas and hyperbolas are not at all what we have in mind.

The best way to see what graphs—or at least diagrams of graphs—are is to look at Figure 8.1 and subsequent figures in this chapter. Having thumbed through the chapter, you probably have a good intuitive idea of what a graph is. Whatever the type of graph G, there are always a nonempty set V_G of elements called **vertices** of G and a set E_G of elements called **edges** of G. We will deal only with finite graphs, meaning the sets V_G and E_G are finite. In the diagrams, the vertices are represented by points and the edges are represented by segments or any other curve connecting the vertices. If G is a **simple graph**, then we can say that the set of edges is a subset of all the unordered pairs of vertices. So, given any 2 different

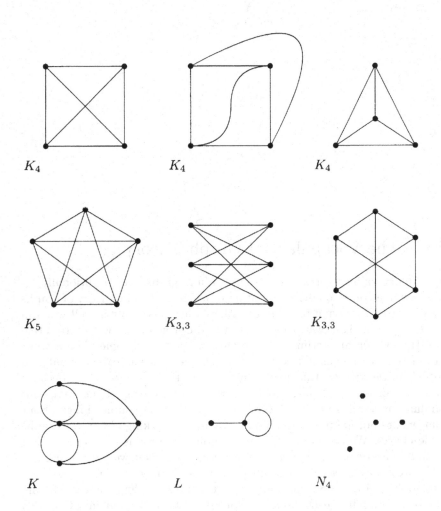

FIGURE 8.1. Diagrams of Graphs.

vertices x and y in V_G, then the set $\{x, y\}$ either is in E_G or is not. If $\{x, y\}$ is in E_G, then we say that the edge $\{x, y\}$ has **ends** x and y, that the edge $\{x, y\}$ **joins** x and y, that x and y are **adjacent**, that x and y are **neighbors**, and that the edge $\{x, y\}$ is **incident** to x and to y. Edges sharing a vertex are also said to be **adjacent**. None of this language should be surprising.

If, for simple graph G, we have $|V_G| = n$ and $|E_G| = \binom{n}{2}$, meaning there are n vertices and that all $\binom{n}{2}$ possible edges are edges of G, then we call G a **complete graph** and denote this graph by K_n. See diagrams for K_4 and K_5 in Figure 8.1. In the figure, graph K_4 is represented by 3 different diagrams. In the first of these, the 4 vertices are represented by the 4 corners of a square and the 6 edges are represented by the 4 sides and 2 diagonals of the square. That the diagonals of the square intersect in the plane has nothing to do with the complete graph consisting of 4 vertices and 6 edges. Also, it does not matter which curves we use to represent an edge joining 2 vertices, although common sense tells us that no curve representing an edge should be drawn to pass through a point representing a vertex other than its ends. For example, each of the following 5 figures is a diagram for K_4.

Our language for graphs easily slides over into that describing the diagrams. Formally, graph K_4 is given by

$$V_G = \{p, q, r, s\} \text{ and } E_G = \{\{p, q\}, \{p, r\}, \{p, s\}, \{q, r\}, \{q, s\}, \{r, s\}\}$$

However, graphs are frequently "given" by their diagrams, because the diagrams make so much more information visually available. The diagram ⊠ conveys information more easily than the display above, even in this very simple example. We already have enough language to understand that it might be exceedingly difficult to tell whether 2 given diagrams or graphs are actually "the same" graph. Simple graphs G and G' are **isomorphic** if there is a one-to-one correspondence $x \to x'$ between the set V_G of vertices of G and the set $V_{G'}$ of vertices of G' and if there is a one-to-one correspondence between the set E_G of edges of G and the set $E_{G'}$ of edges of G' such that $\{x, y\}$ is an edge of G iff $\{x', y'\}$ is an edge of G'. Any 2 complete graphs having the same number of vertices are isomorphic. A simple graph usually has many diagrams. In general, it is very, very difficult to tell whether 2 graphs are isomorphic or not.

Graphs G and G' are certainly not isomorphic unless $|V_G| = |V_{G'}|$ and $|E_G| = |E_{G'}|$. Although each of these 2 conditions holds for the graphs in Figure 8.2, these 2 graphs are essentially different. In Figure 8.2, we see (knowing nothing about organic chemistry, we ignore the caption) that in each of the graphs there are 4 vertices (carbon) that are adjacent to 4 other

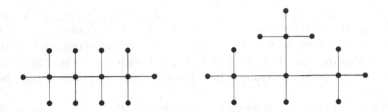

FIGURE 8.2. Butane and Isobutane.

vertices and that there are 10 other vertices (hydrogen) that are adjacent to only 1 other vertex. We observe that in the graph on the right there is a vertex that is adjacent to 3 vertices which are, in turn, adjacent to 4 vertices, while the graph on the left has no such vertex. We need more language to more easily express this situation. If a vertex v is adjacent to exactly $d(v)$ edges, then we say v has **degree** $d(v)$. Thus, each of the 2 graphs in Figure 8.2 has 4 vertices of degree 4 and 10 vertices of degree 1. However, since the graph on the right has a vertex that is adjacent to 3 vertices of degree 4, while no vertex of the graph on the left has this property, the graphs cannot be isomorphic.

If the degree $d(v)$ of vertex v is even, then v is said to be **even**; if the degree $d(v)$ of vertex v is odd, then v is said to be **odd**.

Observation. For a simple graph G,

$$\sum_{v \in V_G} d(v) = 2|E_G|.$$

In words, the observation says that if we sum the degrees of all the vertices, then we get twice the number of edges. This must be so because each edge has 2 ends. Further, since $2|E_G|$ is obviously even, then the left side of the equation above must be even. So, since the sum of an odd number of odd integers is odd, the number of odd terms added on the left must be even.

Corollary. Every simple graph has an even number of odd vertices.

A simple graph is **planar** if the graph has a diagram that can be drawn in the plane in such a way that no 2 of the curves representing the edges intersect, except at vertices. For example, K_4 is planar as seen by 2 of the 3 diagrams representing the graph in the top row of Figure 8.1. The graph having the diagram on the left in Figure 8.3 is also planar, as can be seen in the 3 steps shown in the figure. It has been shown that each planar graph has a diagram where all its edges are represented by (straight) segments that intersect only at their ends.

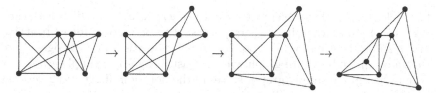

FIGURE 8.3. Diagrams of a Planar Graph.

The simple graph $K_{3,3}$ in the center of Figure 8.1 has a story behind it. We have 3 new houses in a row on the left and 3 utilities in a row on the right (say: power, water, and gas) and the problem is to draw utility lines from each utility to each of the houses without any of these 9 lines crossing. It might be fun to try doing this for a while, and, if you happen to know about something called the Jordan Curve Theorem, convince yourself that it cannot be done. You may even want to try to determine whether K_5 is planar or not. Before we state that neither graph is planar, we want to introduce some more terminology.

Simple graph H is a **subgraph** of simple graph G if V_H is a subset of V_G and if E_H is a subset of E_G. Note that since H is itself a simple graph the ends of an edge of H are automatically in V_H, as well as in V_G. So, a subgraph is just what the name suggests.

A **subdivision** of a simple graph is obtained by inserting (a finite number, possibly 0) vertices of degree 2 into the edges, making the proper replacement of edges indicated by the new diagram. This idea is one of the many in graph theory are made completely clear by a few pictures and whose description in words requires the diagrams in order to understand what is being said anyway. For example, •––•• is a subdivision of •––•, and
is a subdivision of . However, is not a subdivision of .
Is it clear that the planarity of a graph cannot be affected by a subdivision? This leads to a theorem that is neat, although not very useful. We state, without proof, this elegant result, which can be rudely paraphrased, If a graph is not planar, then there is a K_5 or a $K_{3,3}$ hidden in there somewhere.

Kuratowski's Theorem. A simple graph is planar iff the graph has no subgraph isomorphic to a subdivision of K_5 or $K_{3,3}$.

The graph $K_{3,3}$ is an example of a special type of graph. A simple graph G is **bipartite** (rhymes with *kite*; ends in *tight*) if the set V_G of vertices is the union of 2 disjoint sets X and Y such that each edge in G has 1 end in X and 1 end in Y. The sets X and Y are called the **bipartite parts** of G. In $K_{3,3}$, we can take X to be the set of houses, Y is then the set of utilities, and each edge joins a house and a utility. In fact, each house is joined to each utility in this example of a bipartite graph. In general, if X

has r elements, Y has s elements, and each vertex in X is adjacent to each vertex in Y, then the bipartite graph is said to be a **complete bipartite graph** and is denoted by $K_{r,s}$. So, $K_{r,s}$ has $r + s$ vertices and rs edges. Note that a complete bipartite graph with at least 2 edges is a simple graph but is not a complete simple graph, since, in the notation above, no 2 vertices in X are adjacent and no 2 vertices in Y are adjacent.

We might want to consider the definition of the word *graph* to be such that what we have been calling diagrams for simple graphs *are* graphs. This is perfectly reasonable. It is, after all, the diagrams that are driving the show. We could allow the edges to be anything—thinking of curves in the plane or space—but we need something that answers the question, What are the ends of a given edge? This is an easy adjustment. We simply think of a graph G as a triple (V_G, E_G, h_G), where h_G answers the question. Of course, V_G is the set of vertices and E_G is the set of edges, as before, but how should we describe h_G? In order to regain our idea of a simple graph we need h_G to be a mapping from the set of edges to the sets of sets $\{x, y\}$ with x and y distinct vertices. But this is not enough; the mapping h_G must also be one-to-one. That is, if $h_G(e_1) = \{x, y\}$ and $h_G(e_2) = \{x, y\}$, then we must have the edges e_1 and e_2 equal. Now, we are indeed back to our simple graphs, but is this where we now want to be? We have been exposed to a different idea. Why should h_G be one-to-one? Why can't we allow for more than 1 edge to join 2 vertices? If we do make this allowance, such edges are said to be **parallel**. Beware, parallel edges always intersect at their ends, in spite of the conventional use of the word in euclidean geometry. So the graph $\triangle$ has 2 vertices and 2 edges, which are parallel. The bottom left graph, which is labeled K, in Figure 8.1 has 2 pairs of parallel edges. We will get back to this famous graph later. In the definition of a subgraph H of graph G, we now need $V_H \subseteq V_G$, $E_H \subseteq E_G$ and also $h_H(e) = h_G(e)$ for every edge e in E_H.

Well, the floodgate is open. We notice that we have, so far, not allowed the ends of an edge to be equal? Why not? If such an edge is allowed, then the edge is called a **loop**. The graphs $\triangle$ and $\bigcirc$ each have 1 vertex and 1 edge, which is a loop. The bottom-center graph in Figure 8.1 has 2 vertices and 2 edges, 1 of which is a loop. A vertex having no adjacent edge is an **isolated** vertex. The **null graph** N_p has p isolated vertices and no edges. The null graph N_4 is the bottom-right graph in Figure 8.1. (Those who allow a graph not to have any vertices call this graph the **empty graph**.)

Some graph theorists use **multigraph** to describe what we get by allowing parallel edges but not loops. We will follow this usage, knowing that others use the word to allow both parallel edges and loops. (Note that the definition of degree of a vertex will have to be altered if loops are allowed.) And, what does the single word "graph" mean? The problem is that there is no standard usage. Who knows what a *graph* is? Each graph theorist does, taking their lead from a famous character that was created by a mathematician:

> "When I use a word," Humpty Dumpty said in a rather scornful tone, "it means just what I choose it to mean—neither more nor less."

One definition that is not in question is that of the n-**cube**, which is denoted by Q_n. For nonnegative integer n, the n-cube Q_n is the graph whose vertices are the n-digit binary sequences, with 2 of the vertices adjacent iff their sequences differ in exactly 1 position. Since Q_0 consists of only 1 vertex, which is the empty sequence, and no edges, then Q_0, Q_1, and Q_2 have the diagrams $\bullet$, $\bullet\!\!-\!\!\bullet$, and $\square$, respectively. One of the exercises below and some in later sections explore the properties of Q_n.

Homework Graphs 1.

1. Give diagrams for the 11 nonisomorphic graphs having 4 vertices.

2. Show that the graphs with the diagrams labeled $K_{3,3}$ in the second row of Figure 8.1 are actually isomorphic.

3. Prove that the number of people from Kansas who have met an odd number of other people from Kansas is even.

4. Answer the following questions concerning n-cubes Q_n.

   (a) Draw a diagram of the 3-cube with the vertices labeled as the sequences.
   (b) How many vertices does Q_n have?
   (c) What is the degree of each vertex of Q_n?
   (d) How many edges does Q_n have?
   (e) Show that Q_n is bipartite.

5. Show that there are $2^{\binom{n-1}{2}}$ simple graphs having $\{1, 2, 3, \ldots, n\}$ as its set of vertices and such that each vertex has even degree. (Note that the number itself should suggest a method of proof. The number is $2^{\binom{n-1}{2}}$ and not $2 \cdot \binom{n-1}{2}$.)

6. Is the graph below planar?

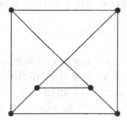

7. How many ways can 7 students go on vacation with each sending a postcard to 3 of the others and each receiving a post card from precisely the 3 to whom they sent cards? (Substantiate your answer.)

8. Suppose bipartite graph G has bipartite parts X and Y. A **complete matching from** X **to** Y is a one-to-one correspondence between the vertices of X and a subset of the vertices of Y such that corresponding vertices are adjacent. Use Hall's Marriage Theorem to show that if the degree of each vertex in X is greater than or equal to the degree of each vertex in Y, then there is a complete matching from X to Y.

§52. Walks, Trails, Circuits, Paths, and Cycles

Throughout this section, we suppose G is a simple graph (V_G, E_G).

We begin with a lot of language, most of which describes Figure 8.4.

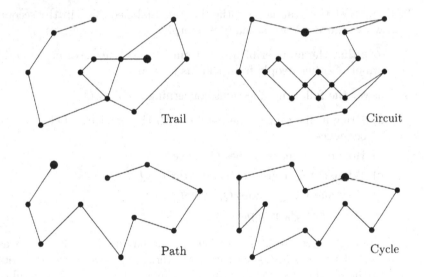

FIGURE 8.4. Special Walks.

A **walk** in G of **length** k, for $k > 0$, is a sequence

$$v_0, e_1, v_1, e_2, v_2, e_3, v_3, \ldots, v_{k-1}, e_k, v_k$$

of length $2k+1$ that alternates between vertices of G and edges of G, whose initial term is a vertex v_0 called the **initial end**, whose terminal term is a vertex v_k called the **terminal end**, and is such that for $1 \leq i \leq k$ we have $e_i = \{v_{i-1} v_i\}$. This walk is also called a **walk from** v_0 **to** v_k. It requires only a little imagination to picture ourself walking along the edges of a graph. A walk in G is **closed** if its ends are equal. (So, a closed walk brings us back to where we started.) A **trail** in G is a walk in G whose edges are distinct. A **circuit** in G is a trail in G that is a closed walk in G. (We get back to our starting point without traversing any edge twice.) A **path** in G is a trail in G with distinct vertices. So, a path is a walk with distinct vertices and distinct edges. A **cycle** in G is a circuit in G such that no 2 vertices other than the ends are equal. A cycle of length k has k distinct vertices and k distinct edges, is called k-**cycle**, and is denoted by C_k. Again, we warn the reader that the terminology is not standardized.

A path is **maximal** if it is not possible to insert vertices and edges into the sequence to obtain a longer path. Since our graphs are finite, there are only a finite number of edges and so every path is itself maximal or can be extended to a maximal path. We can define $d_G(u, v)$, the **distance** from vertex u to vertex v to be 0 if $u = v$, the length of a shortest path from u to v if such a path exists, and, otherwise, to be ∞ if no such path exists. If there is a path from each vertex to each other vertex, then we say that G is **connected**, otherwise the graph is **disconnected** and is made up of several connected **components**. In a connected graph there are no vertices u, v in G for which $d_G(u, v) = \infty$.

Suppose every vertex in G is of degree at least 2. There is a maximal path in G with initial end, say v. Since the degree of end v is at least 2, then v has a neighbor on the path and has another neighbor w that is not on the path. Let e be the edge in G that joins neighboring vertices v and w. However, since we cannot extend that path to reach any new vertex from v because the path is maximal, then w must already be on the path. Therefore, we can attach the sequence e, v to the end of that part of the maximal path that is a path from v to w and so obtain a cycle.

Observation. If each vertex of a simple graph has degree at least 2, then the graph has a cycle.

An **euler** walk, trail, circuit, path, cycle in G is, respectively, a walk, trail, circuit, path, cycle in graph G that contains all the edges and all the vertices. (Some graph theorists prefer "Eulerian" as the adjective, but, in any case, Euler's Swiss name is pronounce *oiler* in English. Euler is the first modern giant of mathematics.) A graph having an euler trail must be connected, because an euler trail contains all the vertices. In an euler trail, each edge appears exactly once but a vertex may appear more than once.

Suppose a connected simple graph G has an euler circuit. Identifying the initial points in our mind, we see that the edges adjacent to a vertex come in pairs as determined by the euler circuit. Any circuit leading into a vertex by an adjacent edge must also depart to an adjacent vertex by another adjacent edge. Since all the edges appear in the euler circuit, then there are an even number of edges adjacent to any vertex on the euler circuit. Since all the vertices appear in an euler circuit, every vertex of G is even.

Conversely, now suppose that each vertex of connected simple graph G with at least 2 vertices has all vertices of even degree. We intend to form an euler circuit. We begin with an arbitrary vertex v and arbitrarily form a trail from v until we get back to v, forming a circuit. This is not only possible but mandatory; since each vertex is of even degree our departure from any vertex in this process is assured if we can get there in the first place, no matter how many times we have already passed through the vertex before. (We "use" exactly 2 edges every time we pass through a vertex.) Since the graph has a finite number of edges, we eventually have to get back to v. We have made a circuit, and, if we have been exceptionally lucky, we have an euler circuit. Chances are, however, we have not traversed all the edges, as required. In this case, we need to arrange some side trips to our grand tour. Suppose some vertex u that we have already visited has an edge, and, hence, an even number of them, that have not yet been traversed. We start out from u on a trail that traverses only edges that have not previously been traversed. We must eventually get back to u to complete a circuit. (Even if we pass through v, we keep going until we get back to u, which we must be able to do.) Now, we simply insert at u this circuit into our first circuit from v to form a longer circuit from v. If our luck has been bad and we still have edges not traversed, then we play the side trip game again, and again, and again, until there are no more edges left with which to play. Since G is connected, we must have a circuit that traverses all the edges and all the vertices. We have proved the following theorem. The proof of the corollary is left as an exercise, with the hint that a temporary edge might bridge the odd vertices.

Theorem. A connected simple graph with at least 2 vertices has an euler circuit iff all vertices are of even degree.

Corollary. A connected simple graph has an euler trail but no euler circuit iff there are exactly 2 odd vertices.

One way to think of this theorem and its corollary is that together they tell us when we can trace a diagram with a pencil while traversing each edge exactly once but never lifting the pencil off the paper.

It is not often that we can give an exact date when a branch of mathematics began. However, graph theory is an exception. In 1736, Leonhard Euler published a paper that grew out of the now famous Königsberg bridge problem. At the time there were 7 bridges over the branches of the Pregel

River that connected the 4 sections of the city, as shown in Figure 8.5. Supposedly the citizens spent Sundays trying to find a way to walk about the city by crossing each of the 7 bridges exactly once. They must have eventually realized that they could not do so. Anyway, Euler represented the bridges by edges and bits of land by vertices and graph theory was born. The result of Euler's model is the multigraph labeled K in Figure 8.1. Euler proved, in particular, the theorem above and its corollary.

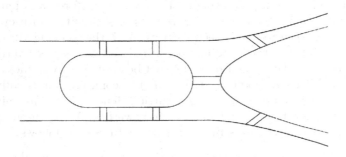

FIGURE 8.5. The Seven Bridges of Königsberg

A **hamilton** path in G is a path in G that contains each vertex exactly once; a **hamilton** cycle in G is a cycle in G that contains each vertex exactly once, with the ends identified as 1 vertex. So, a graph having a hamilton path or a hamilton cycle must be connected. Although each vertex appears in a hamilton path and in a hamilton cycle, there may be edges that do not appear in such a sequence. The hamilton paths are named after Sir Rowan Hamilton, who proposed the puzzle of finding what we call a hamilton cycle along the edges of a dodecahedron that visits all the 20 cities (say, starting from and returning to Dublin) that named the 20 vertices of the dodecahedron. The puzzle is included in the exercises and might best be attacked at this time.

Euler circuits visit each edge only once; hamilton cycles visit each vertex only once if we identify the ends. Surprisingly, there seems to be no helpful connection between these concepts. Knowing whether euler circuits exist in a given graph and finding them when they do is not a difficult problem, as we have seen. However, this is not at all the case with hamilton cycles. Even the fastest computers can be swamped by the task of finding a hamilton cycle in a graph.

We respectively define a walk, trail, circuit, path, cycle to be either **odd** or else **even** as its length is either odd or else even. We will examine the relation of among simple graphs being partite, having odd circuits, and having odd cycles.

Suppose that graph G is bipartite. The set of vertices is then the union of disjoint sets X and Y of vertices such that each edge in G has an end in X and another in Y. No vertices in X are adjacent; no vertices in Y are adjacent. Any circuit or cycle must alternate between the vertices of X and Y and so have an equal number of vertices in each of X and Y when we identify the equal ends. Hence, each circuit (if there are any) and each cycle (if there are any) must have even length. The same argument with a bit more color requires painting all the vertices in X red and painting all the vertices in Y green. Then the vertices of a circuit or of a cycle must alternate between red and green vertices. With the ends identified, we see that there must be the same number of red vertices as green vertices in a circuit or cycle. The circuit or cycle must be even. That was the easy part.

Suppose G has no odd cycle. If G is not connected, it is sufficient to show that each component of G is bipartite. Hence, we may as well assume that G is connected. Let v be an arbitrary vertex. We partition the set of vertices into 2 disjoint sets by their distance from v, as follows.

$$X = \{x \mid d(v, x) \text{ is even }\} \quad \text{and} \quad Y = \{y \mid d(v, y) \text{ is odd }\}.$$

If we show that no 2 vertices of X are adjacent and that no 2 vertices of Y are adjacent, then we have proved that the graph is bipartite. Suppose that vertices u and w are distinct vertices in the same set, either X or Y. Thus, a shortest path P from v to u and shortest path Q from v to w are either both of odd length or both of even length. Let v' be the last vertex that P and Q have in common. See Figure 8.6. If $v' \neq v$, then the

FIGURE 8.6. Shortest Paths from v to u and to w.

parts of paths P and Q that are paths from v to v' must have the same length, since since neither can be shorter than the other. Whether $v' = v$ or not, this means that the path form v' to u and the path from v' to w are either both of odd length or else both of even length. Further, v' is the only vertex that appears in both of these paths. We concatenate the first path written backwards (and so is a shortest path from u to v') and the second path (from v' to w) to obtain a path of even length from u to w. Attaching the edge joining u and w, if it existed, would create a cycle of odd length. Since, by our hypothesis such a cycle does not exist, vertices u and w cannot be adjacent, as desired. Therefore, G is bipartite.

If we assume that G contains no odd circuit, we have only to simplify the proof above by not worrying about v'. Since a circuit can have repeating vertices, we need only concatenate a shortest path from u to v and a shortest path from v to w to get a trail of even length from u to w. Assuming u and w adjacent would imply the existence of an odd circuit. Again, G is bipartite.

Since we have shown that G is bipartite iff G has no odd cycle and that G is bipartite iff G has no odd circuit, it follows that G has no odd circuit iff G has no odd cycle. We throw all this together as a theorem.

Theorem. For a simple graph G, the following are equivalent.

1. G is bipartite.

2. G has no odd circuit.

3. G has no odd cycle.

From the diagrams, it seems clear that a circuit contains (as a subsequence) a cycle. To give an argument for this, we suppose simple graph G has a circuit C, which is a trail from v to v for some vertex v. (The edges of a trail are distinct. If the vertices of C are also distinct with the ends identified as 1 vertex, then we have a cycle and we are done.) Circuit C contains a (there may be more than 1) shortest circuit S from w to w where w is a term in C. Circuit S exists because we know there is at least 1 circuit from v to v and because there are only a finite number of vertices. Except for the initial end w, no vertex term in S appears more than once, as otherwise there is a shorter circuit. So, this circuit is a cycle.

Theorem. In a simple graph, each circuit contains a cycle.

The converse of this theorem is true and trivial.

Some comments about notation and definitions are in order. Since a simple graph has no loops or parallel edges, the notation for a walk can be greatly simplified by ignoring the edges without any possible confusion. So, the 3-cycle $a\{a, b\}b\{b, c\}c\{c, a\}a$ can just as well be written "$abca$." Note that we have already dropped the commas, using them only when their absence would cause confusion. This convention for simple graphs does make life easier. Further, our cycles presently have distinguished ends. Generally, we want to identify the 6 3-cycles

$abca,$	$bcab,$	$cabc,$
$acba,$	$bacb,$	$cbac,$

as being equivalent. We want to be able to start the cycle with any of its vertices and take either of its adjacent vertices on the cycle as the second term. Hence, in general, we want to identify the $2n$ cycles equivalent to a given n-cycle. Frequently it is necessary to discern whether the word "cycle" refers to a walk (with ends) or to a class of equivalent cycles from the context. One convention is to drop the repeated vertex when referring to an equivalence class. For example, "cycle abc" refers to the class of 6 cycles displayed above.

Similar comments refer to circuits. For an example of the difference, in the proof above that a circuit contains a cycle, we would have the circuit (as a sequence) $abcdefgca$ contain the shortest circuit (as a sequence) $cdefgc$, although as an equivalence class the circuit contains the shorter cycle abc.

We associate an isolated vertex • with the 0-cube Q_0. As the point moves a unit length in a given direction, the point sweeps out a segment, •—•, which we associate with the 1-cube Q_1. As the segment moves a unit length in a perpendicular direction, the segment sweeps out a square ⊓, which we associate with the 2-cube Q_2. As the square moves a unit length in a third perpendicular direction, the square sweeps out a cube , which we associate with the 3-cube Q_3. As the cube moves a unit length in a fourth perpendicular direction, the cube sweeps out a **tesseract**. So, a tesseract is a 4-dimensional figure, which we associate with the 4-cube Q_4. Of course, showing 4 dimensions on a 2-dimensional page has its difficulties. One representation of a tesseract is shown in Figure 8.7.

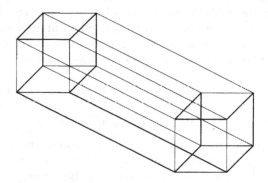

FIGURE 8.7. The Tesseract.

Homework Graphs 2.

1. Prove that a connected simple graph has an euler trail but no euler circuit iff there are exactly 2 odd vertices.

2. Prove that the citizens of Königsberg were doomed in their effort to walk about the city, crossing each of the 7 bridges exactly once.

3. Give examples, where they exist, of simple graphs G such that

    (a) there is an euler circuit and a hamilton cycle.

    (b) there is an euler circuit but no hamilton cycle.

    (c) there is a hamilton cycle but no euler circuit.

    (d) there is neither a hamilton cycle nor an euler circuit.

4. Give a simple graph that has an euler trail but no euler circuit.

5. Show the graph below has a hamilton path but no hamilton cycle.

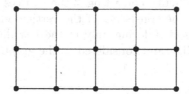

6. Solve Hamilton's original problem about traversing the vertices on a dodecahedron in a cycle, using the following graph as a model of the dodecahedron.

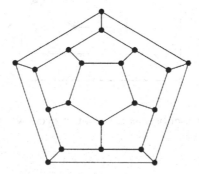

7. Which n-cubes have an euler circuit?

8. A drawing of a tesseract that has more symmetry than Figure 8.7 is given below. (It is easier to find the 8 cubes that are the "faces" of the tesseract in this drawing.) Find an euler circuit in this diagram.

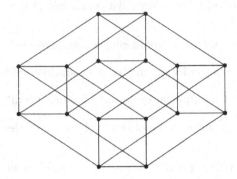

9. Each different representation of Q_4 provides further insight into the 4-dimensional tesseract. An informative diagram of a tesseract is given below. (Think of the fourth dimension as "out." So, the inner cube is the 3-cube consisting of the vertices ending in 0, the outer cube is the 3-cube consisting of the vertices ending in 1, and edges join these vertices differing only in the last digit.) Find a hamilton circuit in the diagram, considered as the graph Q_4.

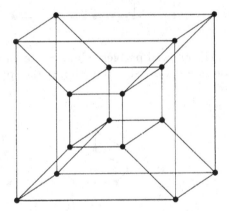

10. Show that the n-cube has a hamilton cycle when $n > 1$.

§53. Trees

Throughout this section we suppose G is a simple graph (V_G, E_G).

A **tree** is a connected simple graph that has no cycles. See Figure 8.8. A vertex of degree 1 in a tree is called a **leaf**. If the components of a disconnected simple graph are trees, that is, if the disconnected graph has no cycles, then the graph is called a **forest**. A **spanning subgraph** of graph G is a subgraph of G that contains all the vertices of G. A **spanning tree** of graph G is a tree that is a spanning subgraph of G.

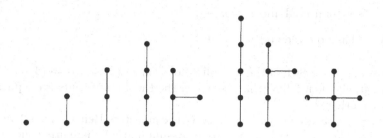

FIGURE 8.8. The Smallest Trees.

We begin our short study of trees with an easy observation. Since a tree is connected, the degree of each vertex of a tree is positive. Since we know that a graph for which each vertex has degree greater than 1 must have a cycle, then a tree must have at least 1 vertex of degree less than 2. Suppose a tree has at least 2 vertices and has a vertex v as a leaf. Consider a maximal path with end v, say from v to w. So, w is adjacent to a vertex on the maximal path, but w cannot be adjacent to any other vertex on the maximal path, as otherwise there would be a circuit and hence a cycle. Further, w cannot be adjacent to any vertex not on the maximal path since the path is maximal. Thus the degree of w is 1, and w is a leaf.

Theorem. A tree with at least 2 vertices has at least 2 leaves.

We next show that a simple graph G is connected iff G has a spanning tree. If G has a spanning tree then the graph is clearly connected since the tree is connected and contains all the vertices. Conversely, suppose G is connected and not a tree. Then G contains a cycle. We form a graph G_1 by deleting only 1 edge of this cycle from G. Graph G_1 is connected. If G_1 is not a tree, then there is a cycle in G_1, which is also a cycle in G. We form a graph G_2 by deleting only 1 edge of this cycle from G_1. Graph G_2 is connected. If G_2 is not a tree, then there is a cycle in G_2, which is also a cycle in G. We continue in this fashion, obtaining a connected graph at each step, until we reach a connected graph that has no cycles, a tree. Since all the points of G are in this tree, the tree is a spanning tree. We have proved the first of the following 2 theorems.

Theorem. A simple graph G is connected iff G has a spanning tree.

Theorem. If G is a simple graph having n vertices with $n \geq 2$ and having m edges, then the following are equivalent.

1. G is a tree.

2. For each 2 vertices v and w, there is a unique path from v to w.

3. G is connected and $n = m + 1$.

4. G has no cycle and $n = m + 1$.

We will prove each of the 4 implications $(1) \Rightarrow (2) \Rightarrow (3) \Rightarrow (4) \Rightarrow (1)$, in turn. It will then follow that each 2 of the the 4 statements are equivalent to each other.

$(1) \Rightarrow (2)$. Since G is a tree, then G is connected. Hence, given 2 vertices v and w, there is a path from v to w. Assuming that there are 2 such paths implies there is a cycle, as follows. Forming the walks from the first vertex after which the 2 paths differ to the next vertex that the remaining parts of the 2 paths have in common (which could be w), we have 2 paths with the same ends but are otherwise disjoint. Concatenating one of these with the reverse of the other produces a cycle in G, a contradiction. So, the path from v to w is unique.

$(2) \Rightarrow (3)$. First, the existence of the paths means that G is connected. We will prove the desired formula by mathematical induction on n, the number of vertices. For the basis step, if $n = 2$, there is 1 edge and the result is trivial: $2 = 1 + 1$. For the induction step, we assume that $(2) \Rightarrow$ (3) for graphs having k vertices for some k with $k \geq 2$ and then prove the implication for graphs having $k + 1$ vertices. Suppose G is a graph with $k + 1$ vertices and m edges. Since $d(v) \geq 2$ for each vertex would imply the existence of a cycle, which would contradict the uniqueness of paths between each 2 vertices, then there must be at least 1 vertex in G, say w, of degree 1. Consider the graph H formed by deleting w and its adjacent edge from G. So H is a graph with k vertices and $m - 1$ edges and is such that there is a path from each vertex to any other vertex in H. By the induction hypothesis, then $k = (m - 1) + 1$. So, $(k + 1) = (m) + 1$, as desired. The implication follows by mathematical induction.

$(3) \Rightarrow (4)$. We must show that if G is connected and $n = m + 1$, then G has no cycles. Again, we use mathematical induction on n. If $n = 2$, then there are no cycles and the basis step is trivial. For the induction step, we assume $(3) \Rightarrow (4)$ if $n = k$ for some k with $k \geq 2$ and then prove the implication if $n = k + 1$. So, suppose G has $k + 1$ vertices and m edges with $k + 1 = m + 1$. Now, $d(v) \geq 2$ for each vertex would imply at least $k + 1$ edges, since the sum of all degrees is twice the number of edges. So, there is at least 1 vertex, say w, of degree 1. Consider the graph H formed by deleting w and its adjacent edge from G. So H is a connected graph with k vertices and $k - 1$ edges and does not have any cycles, by the induction hypothesis. Any cycle in G must then contain w. However, this is impossible since w has degree only 1 in G. Hence, G has no cycles, as desired. The implication follows by mathematical induction.

(4) ⇒ (1). We must show that G is connected. Again, we use mathematical induction on n. Again the basis step is easy. For $n = 2$, we have $m = n - 1 = 1$. So G is a tree consisting of 2 vertices and 1 edge joining them. For the induction step, we assume (4) ⇒ (1) for graphs with k vertices for some k with $k \geq 2$ and let G be a graph with $k + 1$ vertices and k edges. As in the previous argument, there must be a vertex w such that $d(w) < 2$. If G has no vertex of degree 1, then by deleting all the vertices of degree 0, we have a subgraph with all degrees at least 2 and so must have a cycle, contradicting (4). Thus we may suppose $d(w) = 1$. Graph H formed by deleting w and its adjacent edge from G, has no cycles, has $k - 1$ edges, and, by the induction hypothesis, is connected. In this case, then G must also be connected, as desired. The implication follows by mathematical induction. This finishes the long proof of the equivalence of the 4 properties in the theorem stated above.

Theorem. Every tree is bipartite.

Our proof uses mathematical induction on the number n of vertices in a tree. We need to show that the set of vertices is the union of disjoint sets X and Y, where each edge in the tree has 1 end in X and 1 end in Y. For the basis step, we consider the tree with 1 vertex v. Here $X = \{v\}$ and $Y = \emptyset$. (You may want to start with trees having at least 1 edge; for the case $n = 2$, each of X and Y contains 1 vertex.) For the induction step, we assume the proposition is true for trees having k vertices and prove that the proposition is true for trees having $k + 1$ vertices when $k \geq 1$. Suppose tree T has $k + 1$ vertices and that z is a leaf of T adjacent to edge t and vertex y. We form tree H by deleting exactly z and t from T. By the induction hypothesis, there are disjoint sets X' and Y' whose union is the set of vertices in H with each edge in H having ends in both X' and Y'. We may suppose that y is in Y' without loss of generality. Then $X = X' \cup \{z\}$ and $Y = Y'$ show that T is bipartite, since the only new edge is $\{z, y\}$, which has 1 end in X and 1 end in Y. So, that the proposition holds for the case of k vertices implies that the proposition holds for the case of $k + 1$ vertices. Hence, the proposition holds for all trees by mathematical induction.

Cayley's Theorem. The number of spanning trees of the complete graph K_n with vertices $\{1, 2, 3, \ldots, n\}$ is n^{n-2}.

If n is 1, 2, or 3, then Cayley's Theorem is obvious. We need $n > 1$ for the following argument but can suppose $n > 2$ to make things easier. We develop a one-to-one correspondence between the spanning trees and the set of sequences of length $n - 2$ whose terms are vertices. The sequence is called the **Prüfer code** for the spanning tree.

Given a spanning tree T, we calculate the Prüfer code $v_1 v_2 \ldots v_{n-2}$ as follows. Let l_1 be the leaf that is the smallest integer; let v_1 be the vertex

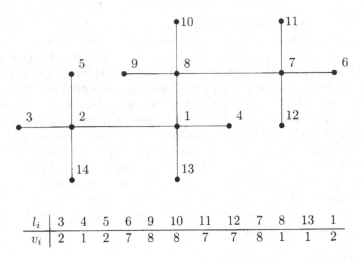

l_i	3	4	5	6	9	10	11	12	7	8	13	1
v_i	2	1	2	7	8	8	7	7	8	1	1	2

FIGURE 8.9. Tree with Prüfer Code 212788778112.

adjacent to l_1. (Remember that the vertices *are* integers.) Delete from T the vertex l_1 and the edge adjacent to l_1, producing a subtree T_1. Let l_2 be the leaf of T_1 that is the smallest integer; let v_2 be the vertex adjacent to l_2. Delete from T_1 the vertex l_2 and the edge adjacent to l_2, producing a subtree T_2. We continue in this fashion until only 2 vertices and the 1 edge joining these 2 vertices remain; 1 of these 2 remaining vertices is the last term added to our sequence of length $n-2$ and the other is the vertex n. For example, the code 2332 is derived for the given tree as follows.

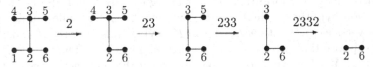

Again, the code is derived by iteratively deleting the leaf with the smallest label and appending the label of its neighbor to the code. After performing $n-2$ iterations, we have a sequence of length $n-2$, with a single edge and its ends remaining. Observe that a vertex that is not a leaf of T is listed once in the sequence for each vertex, except 1, to which it is adjacent. Hence, each vertex of T is listed $d(v)-1$ times in the Prüfer code for T. The leaves of T are absent from the sequence. A more complicated example is given in Figure 8.9.

In order for there to be a one-to-one correspondence, we must be able to recapture the same tree, given the sequence. We do that next.

Given a sequence $v_1 v_2 \ldots v_{n-2}$ of length $n-2$ whose terms are from the first n positive integers, we define a graph having $\{1, 2, \ldots, n\}$ as its set of vertices and its set of edges given as follows. We let L_n be a list of the n vertices. Let l_1 be the smallest integer on list L_n that is not in the given

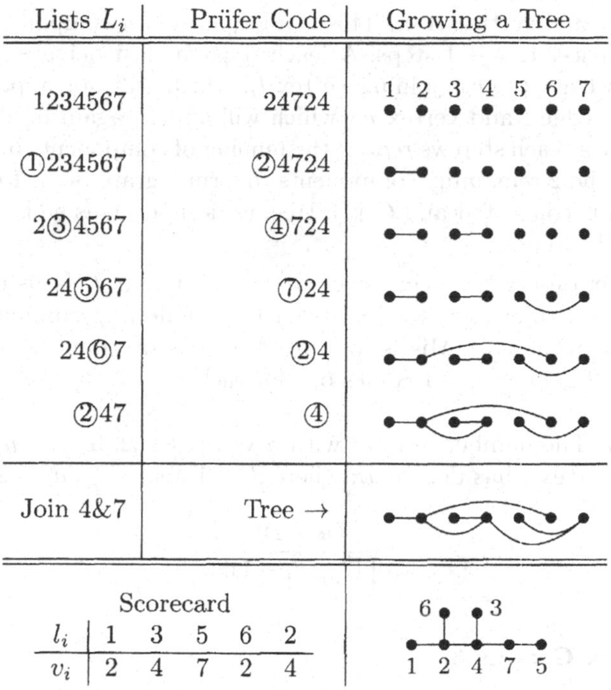

Lists L_i	Prüfer Code	Growing a Tree
1234567	24724	1 2 3 4 5 6 7
①234567	②4724	
2③4567	④724	
24⑤67	⑦24	
24⑥7	②4	
②47	④	
Join 4&7	Tree →	

Scorecard

l_i	1	3	5	6	2
v_i	2	4	7	2	4

6   3

1 2 4 7 5

FIGURE 8.10. Tree for Prüfer Code 24724.

sequence $v_1 v_2 \ldots v_{n-2}$. For our first edge, we take $\{l_1, v_1\}$. We cross l_1 off our list L_n to get a new list L_{n-1}. Let l_2 be the smallest integer on list L_{n-1} that is not in the sequence $v_2 v_3 \ldots v_{n-2}$. For our second edge, we take $\{l_2, v_2\}$. We cross l_2 off our list L_{n-1} to get new list L_{n-2} and continue as before. Our i^{th} edge will be $\{l_i, v_i\}$ for $i = 1, 2, \ldots, n-2$. Our last edge, the $(n-1)^{\text{st}}$ edge, will be the edge having the 2 integers that remain on list L_2 as ends. This process reverses the process of forming the Prüfer code from the tree. For the example in Figure 8.9, given the Prüer code of v_i's, we have an algorithm for finding the l_i's such the the sets $\{l_i, v_i\}$ are the edges. We have recaptured all the edges of the tree.

Let's recap. We started with a spanning tree and developed a sequence. In the previous paragraph, we developed an algorithm that recaptures the tree from the sequence. However, we are not done yet. In order to establish the desired one-to-one correspondence, we must show that if we apply the algorithm to an arbitrary vertex sequence of length $n-2$ then we get a tree. It is evident that the algorithm produces a graph G that has the desired n vertices and that has the desired number $n-1$ of edges. By our theorem on page 169, if G is connected then we do have a spanning tree.

For an example, follow the algorithm of finding the tree for the given sequence 24724 that is shown in Figure 8.10. In general, we begin with the set of n isolated vertices 1–n. This is a graph with n components. We join pairs of vertices in $n-1$ steps. At each step, the last being a special case, we add an edge $\{l_i, v_i\}$ joining vertex l_i, which will not appear again in subsequent edges, and vertex v_i, which will appear again in a subsequent edge. Thus, at each step we reduce the number of components by 1. The last edge joins the 2 remaining components to form a graph with 1 component. The graph is connected and G is a tree, as desired. This ends the proof of Cayley's Theorem.

From our observation that each vertex v of tree T is listed $d(v) - 1$ times in the Prüfer code for T, we have the following corollary given by the solution to an easy Mississippi problem. We are counting sequences of length $n-2$ having $d_i - 1$ copies of i for each i.

Corollary. The number of trees with n vertices 1, 2, 3, ..., n with $n \geq 2$ such that vertex i has degree d_i, where $d_i \geq 1$ and $\sum_{i=1}^{n} d_i = 2(n-1)$, is

$$\frac{(n-2)!}{\prod_{i=1}^{n}(d_i - 1)!}.$$

Homework Graphs 3.

1. Give diagrams for the nonisomorphic trees that have 6 vertices.

2. Find all nonisomorphic spanning trees of $K_{3,3}$.

3. The serious study of trees began with Arthur Cayley's study of the different isomers of C_nH_{2n+2}. In tree language for the case $n = 6$, give diagrams for all the nonisomorphic trees having 14 leaves and 6 vertices of degree 4.

4. Give all the spanning trees for K_4.

5. Find the Prüfer code for each of the following trees.

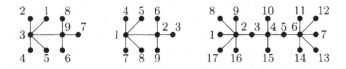

6. Find the trees having the following Prüfer codes.

$$666666 \qquad 123123 \qquad 9191919$$

7. Show that $K_{2,n}$ has $n2^{n-1}$ spanning trees. (This is a special case of the theorem: $K_{m,n}$ has $m^{n-1}n^{m-1}$ spanning trees.)

§54. Degree Sequences

Sequence $d_1, d_2, \ldots, d_n$ of nonnegative integers is **graphic** if there is a simple graph G with n vertices v_i such that $d(v_i) = d_i$ for $i = 1, 2, \ldots, n$. The **degree sequence** for simple graph G is the unique sequence formed by arranging a graphic sequence for G in nondecreasing order. So, if the sequence $d_1, d_2, \ldots, d_n$ is a degree sequence, then we know that the sequence is graphic and that $d_1 \geq d_2 \geq \ldots \geq d_n$. Again, these definitions are not standard.

The theorem that we prove below provides an answer to the question of whether a given sequence is graphic or not. The theorem provides an algorithm that produces shorter and shorter sequences until we get to a sequence we recognize as graphic or not. The original sequence is graphic iff the derived sequences are graphic. For example, we might end up with 11110, which we recognize as the degree sequence for the graph ⋯. Sequence 22211 is both the degree sequence for the tree ⋯ and the degree sequence for the disconnected graph ⋯. The sequence 222222 is graphic for the hexagon ⋯ and for a graph ⋯, consisting of 2 triangles as its components. Although a graph has a unique degree sequence, a degree sequence can be graphic for nonisomorphic graphs.

On the other hand, none of sequences 622222, 54322, and 44440 can be a degree sequence $d_1, d_2, \ldots, d_n$. We can exclude the first 2 of these because they violate the condition $d_1 < n$. (A vertex can be adjacent to at most $n-1$ vertices if there only $n-1$ other vertices.) The third of these sequences is not a degree sequence for essentially the same reason, since we can ignore 0's, which must correspond to to isolated vertices, in determining whether a sequence is graphic or not. Using a different criterion, we see immediately that neither sequence 544321 nor sequence 33311 can be graphic. This follows because we know that the number of odd vertices in a simple graph must be even.

Theorem. If the sequence

$$r, s_1, s_2, \ldots, s_r, t_1, t_2, \ldots, t_p$$

is a degree sequence, then the sequence

$$s_1 - 1, s_2 - 1, \ldots, s_r - 1, t_1, t_2, \ldots, t_p$$

is graphic. Conversely, if the second sequence is graphic, then the first sequence is graphic.

Proving the converse is the easy part of the proof of the theorem. If the second sequence is graphic for simple graph G, then we construct simple graph H by adding to G a new vertex z and r edges such that z is adjacent to precisely each of the r vertices in G having degree $s_i - 1$ for $i = 1, 2, \ldots, r$.

The first sequence is then graphic for H. (The first sequence may not be a degree sequence since the terms may not be in the right order. However, if the second sequence is a degree sequence and $r \geq s_1$, then the first sequence is a degree sequence.)

To prove the first part of the theorem, now suppose that the first sequence is the degree sequence for graph H_0. Let u, v_i, and w_j be the corresponding vertices such that $d(u) = r$, $d(v_i) = s_i$, and $d(w_j) = t_j$. We want to construct a graph G such that the second sequence is graphic for G. If vertex u is adjacent to each of the r vertices v_i, then this is our lucky day. In this case, we construct graph G by deleting from H_0 the vertex u and each edge adjacent to u. The second sequence is then graphic for G. (The graphic sequence may need to be rearranged to form the degree sequence for G.)

We now assume that we were unlucky enough to have some h with $1 \leq h \leq r$ such that vertex u is not adjacent to vertex v_h but is adjacent to to vertex w_k for some k. So, h and k are fixed. We want eventually to get to a graph having the same vertices and the same degree sequence but with u adjacent to each v_i. Since the terms of the degree sequence are nondecreasing, then $s_h \geq t_k$. In case $s_h = t_k$, we form graph H_1 simply by interchanging v_h and w_k in H_0. Otherwise, we must have $s_h > t_k$. Then, since v_h has more neighbors than w_k, there exists a vertex z that is adjacent to v_h but not adjacent to w_k.

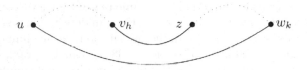

In this case, we replace the edges $\{u, w_k\}$ and $\{v_h, z\}$ by the edges $\{u, v_h\}$ and $\{z, w_k\}$ to obtain a new graph H_1. In either case, graph H_1 has the same degree sequence as H_0 and the same set of vertices. Further, vertex u is now adjacent to vertex v_h, as desired. We repeat this process at most r times until we have graph H with vertex u adjacent to each of the r vertices v_i. Finally, the desired graph G is obtained from H by deleting u and all r edges adjacent to u. The second sequence is graphic for G, and we have finished proving the theorem.

We use sequences 655532 and 444444444 as an illustration. We apply the algorithm of the theorem, where we have underlined the "s_i" below. For the right example only, we must rearrange the derived sequences to nondecreasing sequences before applying the algorithm again.

$$
\begin{array}{ll}
6\underline{65553}2 & 4444\underline{44}444 \\
5\underline{44421} & 33334444 \\
33310 & 4444\underline{3}333 \\
2200 & 3332333 \\
 & 33\underline{333}32 \\
 & 222332 \\
 & 33\underline{2222} \\
 & 21122 \\
 & 22211
\end{array}
$$

Since 2200 is not graphic, then 655532 is not a degree sequence. However, since 22211 is graphic, the sequence of 9 4's is a degree sequence. By working our way up the column on the right, it is very easy to produce a graph having 9 4's as its degree sequence.

Homework Graphs 4.

1. Show that any simple graph has at least 2 vertices of the same degree.

2. Show that neither 6664422 nor 66665421 is graphic.

3. Give a diagram for a graph having degree sequence 6664433.

4. Show that 544421 is not graphic. Find another nondecreasing sequence $d_1, d_2, \ldots, d_6$ of positive integers such that the number of odd terms is even, $d_1 < 6$, but the sequence is not graphic.

5. For $n > 4$, show that there are graphs with n vertices, each of degree 4.

6. If $n > 2$, show that there are graphs with n vertices, each of degree 2.

7. If $k > 1$, show that there are graphs with $2k$ vertices, each of degree 3.

§55. Euler's Formula

Throughout this section we suppose G is a planar graph (V_G, E_G). Recall that a planar graph is a simple graph. Further, when we mention a diagram for a planar graph we suppose that the diagram is one with no edge crossings. The existence of such a diagram is, after all, the distinguishing property that makes a simple graph planar.

Euler is arguably the greatest mathematician of modern times and is certainly the most productive of all time. We can never be sure what is

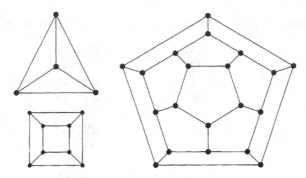

FIGURE 8.11. Planar Graphs for Three Solids.

meant by "Euler's Theorem" out of context. There are so many of them. Even "Euler's Formula" needs some explaining. We might guess that the latter means the beautiful equation $e^{i\pi} + 1 = 0$ involving 5 fundamental constants. This equation is a special case of Euler's more general formula $e^{ix} = \cos x + i \sin x$. However, we have a totally different result in mind here.

Euler's 1758 Polyhedral Formula states that if a convex polyhedral solid has v vertices, e edges, and f faces, then

$$v - e + f = 2.$$

Check out this formula for a regular tetrahedron, a cube, and a regular dodecahedron. We will prove a slight generalization of this formula in the context of connected planar graphs.

Imagine a convex polyhedral surface that is made from a very stretchable substance. Pierce a hole in one face and stretch the surface flat. The result is a diagram for a connected planar simple graph. For example, see Figure 8.11 where the diagrams represent a tetrahedron, a cube, and a regular dodecahedron. The faces of the polyhedral surface are now represented by connected regions of the plane. The pierced face is represented by an unbounded region that is called *the infinite region*. In considering the regions determined by a diagram for a planar graph, we always include the infinite region as a region. These graphs also have the property that every edge borders 2 regions. This is not required of planar graphs in general. Does it make a difference which face is pierced? More generally, does each diagram for a given connected planar graph have the same number of regions? That would be surprising. However surprising it may be, it turns out to be true. For example, the 2 diagrams in Figure 8.12 are for the same graph. Each diagram does determine 9 regions, as numbered in the figure. (It is not necessary for us to prove here the isomorphism of the graphs determined by the 2 diagrams, although the task is not that hard in this case because the numbering of the regions is not random.)

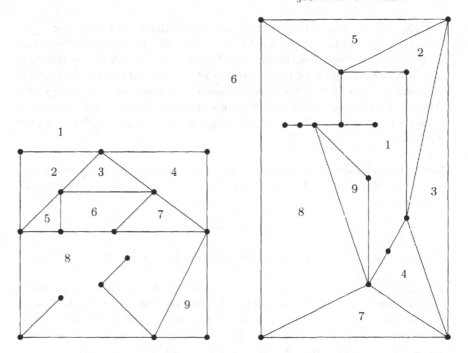

FIGURE 8.12. One Graph, Two Diagrams.

Euler's Polygonal Formula follows if we can show that any diagram with no edge crossings for a connected planar graph G with p vertices and q edges must have r regions where $r = 2 + q - p$. Equation $p - q + r = 2$ is the graph analogue of the polyhedral formula $v - e + f = 2$. The graph ⧉ consisting of 2 triangles shows the necessity of having the given graph connected, since here $p - q + r = 6 - 6 + 3 = 3 \neq 2$.

What happens to the diagram with r regions if we remove an edge from the associated connected planar graph having p vertices, q edges, and n cycles? For an illustration, consider the graph

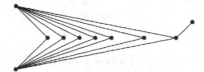

where $p = 10$, $q = 15$, $r = 7$, and $n = \binom{7}{2} = 21$. Removing an edge could result in a disconnected graph. For example, if the "tail edge" is removed from the illustrated graph, then the "tail vertex" is no longer connected to the larger component. If we want to have a connected graph remaining, the edge we remove must be from a cycle. If the edge removed is from a cycle, then the diagram loses 1 region as 2 former adjacent regions melt into 1. Of course, we lose at least 1 cycle. However, we could lose more than

1 cycle, because that edge could be an edge of many cycles. For example, removing any 1 of the 14 edges other than the tail edge from the illustrated graph above, we lose 6 of the former 21 cycles. (Contemplate the difficulty of finding the number of cycles that contain the right edge of the right diagram in Figure 8.12.) We make a final observation before turning to the statement and proof of the graph form of Euler's Formula. The number of cycles for a graph is a function of the graph alone and is independent of any diagram for the graph.

Euler's Formula. If a diagram with no edge crossings for a connected planar graph has p vertices, q edges, and r regions (including the infinite region), then

$$p - q + r = 2.$$

The value of r is independent of the diagram for the graph.

Proof. We prove the following statement using the strong form of mathematical induction on n: Any diagram with no edge crossings for a connected planar graph G having p vertices, q edges, and n cycles has r regions, where $r = 2 - p + q$. For the basis step, if G has no cycles, then the graph is a tree. Since we have $p = q + 1$ and $r = 1$ in this case, then $r = 1 = 2 - (q + 1) + q = 2 - p + q$, proving the formula for the case $n = 0$. For the induction step, we assume that the formula for r holds for connected planar graphs with at most k cycles for some k such that $k \geq 0$. Suppose we have a connected planar graph G with $k + 1$ cycles, p vertices, q edges, and r regions. We form a new graph H by removing exactly 1 edge from G such that this edge is in a cycle of G. This edge must be on a border between 2 regions of G. Thus, H is a connected planar graph with p vertices, $q - 1$ edges, $r - 1$ regions, and at most k cycles. Applying the induction hypothesis to H, we then have $(r - 1) = 2 - p + (q - 1)$. Hence, $r = 2 - p + q$, as desired. So, that the formula holds for such graphs with at most k cycles implies the formula holds for such graphs with $k + 1$ cycles. Therefore, the equation $r = 2 - p + q$ holds for all n by mathematical induction. We rewrite this equation as Euler's Formula $p - q + r = 2$ for the diagram of any connected planar graph. ∎

Several corollaries follow from Euler's Formula. The exercises below ask for proofs of the following 5 corollaries. The first step in proving the first of these is to show that $3r \leq 2q$. As a further hint, Corollaries 2 and 5 are actually corollaries of Corollary 1, and Corollary 4 is a corollary of Corollary 3.

Corollary 1. If a connected planar graph has p vertices and q edges, then

$$q \leq 3p - 6.$$

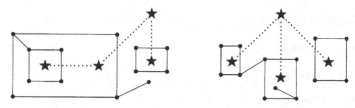

FIGURE 8.13. Different Duals.

Corollary 2. Graph K_5 is not planar.

Corollary 3. If a connected planar graph has p vertices and q edges and if the graph has no triangles, then

$$q \leq 2p - 4.$$

Corollary 4. Graph $K_{3,3}$ is not planar.

Corollary 5. A connected planar graph has a vertex of degree at most 5.

Corollary 5 is the first step in the proof of the 5-color theorem. That is, any map—in the geographic sense of the word—of connected countries can be colored with 5 colors such that no adjacent countries have the same color. This is not terribly hard to prove. Changing "5" to "4," we get the famous 4-color theorem, which is only recently known to be true and is very, very difficult to prove.

A geographic map can be considered as a diagram D for a planar graph G in an obvious way. Associated with D is another graph H, which we call a **dual** of G, defined as follows. The vertices of H are the regions (countries or a possible surrounding sea) of D, and 2 vertices of H are adjacent iff the corresponding countries share a border (not at a vertex only). So coloring the original map (diagram D) with adjacent countries having different colors is the same as coloring the vertices of the dual graph H such that no 2 adjacent vertices have the same color. This is what it means to *color* a graph. (Other definitions for the *dual* of a planar graph produce a planar multigraph, where H has the same number of edges as G.) Some observations are in order. First, a dual is defined only for planar graphs. Note that diagram D determines unique graphs G and H. However, given planar graph G, a diagram for G is not necessarily unique and so a dual of G is not necessarily unique. However, starting with a map and its interpretation as a diagram D for a planar graph G, we see that— even if G is not connected — the dual H is a connected planar graph. For example, in Figure 8.13 we have 2 diagrams (right and left; edges solid) for a disconnected planar graph G. The duals (vertices stared; edges dotted) determined by the diagrams are not isomorphic graphs, although the duals must have the same number of vertices by Euler's Formula. Whole books are written on coloring maps and graphs. We will not go there since our goal is more modest. Chapter 5 is a beginning in this direction.

Homework Graphs 5.

1. Prove Corollary 1.

2. Prove Corollary 2.

3. Prove Corollary 3.

4. Prove Corollary 4.

5. Prove Corollary 5.

Postscript.

Applications of graph theory frequently require **weighted graphs**, where the edges are labeled with numbers, indicating distance, cost, time, etc. There is no known algorithm for finding the most efficient hamilton cycle, where "efficient" means minimizing (or maximizing) the sum of the weights of the edges in the cycle. This is a major unsolved problem in graph theory.

Even more useful for many applications are **digraphs**, short for *directed graphs*. Here we have a set of vertices and an edge is an ordered pair of vertices. So the directed edge (a, b) is distinct from the directed edge (b, a). In a diagram for a digraph, each edge has an arrow to indicate its direction. Digraphs are especially amenable to computation since a digraph can be represented by a square matrix each of whose entries is either $+1$ or 0. If there are n vertices $v_1, \ldots, v_n$, then the graph is given by the n-by-n matrix $[a_{i,j}]$ where $a_{i,j}$ is 1 if (v_i, v_j) is a directed edge and is 0 otherwise. This is a whole new world, which we will not explore.

The Back of the Book

It is hoped that you will wisely use the hints and solutions that are given here. It is foolish to look here immediately after reading a question to see how it is done; you learn next to nothing that way. You will gain more by trying to solve a problem first, even if you get the problem wrong—perhaps especially if you get the problem wrong.

Chapter 1. Elementary Enumeration

§2. Conventions.

Ten Quickies. 1. There are 14 ways to pick 1 person from 14 persons because persons are always distinguishable. **2.** Since pieces of a type of fruit are indistinguishable, there are 2 ways: pick an orange or pick an apple. No, we are not concerned with the manner in which you make your choices; whether you shout, whisper, or write your choice does not concern us. **3.** 3. **4.** 3, since order within the choice was not mentioned. **5.** 15. **6.** 1. **7 & 8.** These are 2 ways of stating the same question and this is a very important principle: Whenever you select r from n distinguishable objects, you are automatically selecting $n - r$ of the objects as well. **9.** 6, which answers the question, How many ways are there to select oranges?

If you select k oranges, then you must take $5-k$ apples. **10.** Since 0 is a possibility for oranges and for apples but not for both, we have $(9+1)(6+1) - 1$ ways.

§3. Permutations.

More Quickies. Since a word is a list, the order of the letters in a word must be taken into consideration. So the word "word" specifically, if implicitly, mentions order. **1.** 5×7. **2.** 26^2. **3.** 26×25. **4.** 21×5. **5.** 3×8. **6.** 5×4; yes, we assume that one person picks 1 chair and the remaining person picks exactly 1 of the remaining unoccupied chairs. **7.** Half as many ways as the answer to #6. **8.** 26^4. **9.** 5×7. **10.** $m \times n$. **11.** 2×6. **12.** $2 \times 6 \times 52$. **13.** 4!. **14.** 13!. **15.** $n!$.

§4. A Discussion Question.

A Discussion Question. The answer probably is not 11; not 36; but 21. Of the students that the author has asked the question on the first day of class over the years, 73% said 36, 26% said 21, but, surprisingly, no one has said 11. See Table 1. The dice in a pair of dice are indistinguishable, although in studying probability we use differently colored dice to find equally likely possibilities. For many games, only the sum matters and so 11 is a reasonable answer. On the other hand, should "doubles" be considered different?

"Sum outcomes" 11 ways (with prob.)	"Visual outcomes" 21 ways (dice are indistinguishable)	"Elementary events" in probability. 36 equally likely ways, with distinguishable dice
2 (1/36)	1&1	(1,1)
3 (2/36)	1&2	(1,2), (2,1)
4 (3/36)	1&3, 2&2	(1,3), (2,2), (3,1)
5 (4/36)	1&4, 2&3	(1,4), (2,3), (3,2), (4,1)
6 (5/36)	1&5, 2&4, 3&3	(1,5), (2,4), (3,3), (4,2), (5,1)
7 (6/36)	1&6, 2&5, 3&4	(1,6), (2,5), (3,4), (4,3), (5,2), (6,1)
8 (5/36)	2&6, 3&5, 4&4	(2,6), (3,5), (4,4), (5,3), (6,2)
9 (4/36)	3&6, 4&5	(3,6), (4,5), (5,4), (6,3)
10 (3/36)	4&6, 5&5	(4,6), (5,5), (6,4)
11 (2/36)	5&6	(5,6), (6,5)
12 (1/36)	6&6	(6,6)

TABLE 1. How Many Ways Can a Pair of Dice Fall?

§5. The Pigeonhole Principle.

Pigeonhole Problems. 6. Every person has from 0 to 19 friends, but 0 and 19 together are impossible. So there are 20 persons (pigeons) and at most 19 possibly different numbers (pigeonholes). **7.** We can partition all such points into 4 classes, where 2 points are in the same class iff their abscissas have the same parity, i.e., are both odd or both even, and their ordinates have the same parity. Since 2 of the 5 points must be in the same class, the midpoint of these 2 points will have integer coordinates. **8.** We have already done this problem. When last seen in the example before this set of problems, this problem was in the disguise of a problem concerning points in the plane. Points correspond to persons and the different colors correspond to the relationships of friends or strangers. Problem solvers often tumble a given problem around to see if restating the problem in a different context throws some light on an attack that leads to a solution.

What is the smallest party of persons for which it is guaranteed that a specific number r of the guests will either all know each other or all will be mutual strangers? This **party problem** seems like a harmless question, does it not? For the case $r = 3$, problem #8 shows that a party of 6 is sufficient and it is easy to argue that a party of 6 is also necessary. It is known that for $r = 4$, a party of 18 is necessary and sufficient. At present, the answers for larger vales of r are not known. It has been conjectured that the solution for the case $r = 5$ won't be found for at least a century. In combinatorics, it is not too difficult to ask questions that no one can answer.

§7. The Round Table.

Homework. 1. $\binom{11}{5}$. **2.** $\binom{52}{5}$. **3.** $\binom{52}{13}$. **4.** $\binom{13}{1}\binom{4}{3}\binom{12}{1}\binom{4}{2}$. **5.** $2^{10} - 1$. We can put YES or NO before each name on the list of friends in 2^{10} ways. The only exclusion is having all NO's. **6.** $12!/[3!4!4!]$. **7.** $(5 \times 7) + (7 \times 4) + (5 \times 4)$ or $\binom{5+7+4}{2} - [\binom{5}{2} + \binom{7}{2} + \binom{4}{2}]$. **8.** $6 \times 8 - 1$. **9.** $21!\binom{22}{5}5!$. **10.** 26×25^9. **11.** $\binom{26}{10} - \binom{17}{10}$ because the number of words consisting of 10 C's and 16 R's that have no 2 C's adjacent is $\binom{16+1}{10}$. **12.** Seat the women 7! ways; select 5 of the 8 spaces into which the men will go in $\binom{8}{5}$ ways; and put the men in those spaces in 5! ways to get a final answer of $7!\binom{8}{5}5!$. **13.** Seat the women 6! ways; select 5 of the 7 spaces into which the men will go in $\binom{7}{5}$ ways; and put the men in those spaces in 5! ways to get a final answer of $6!\binom{7}{5}5!$.

Which of the these questions can you answer now? The first 3 answers are $n!/(n-r)!$, n^r, and $\binom{n}{r}$. It is the fourth question that is troublesome and must be our next principal goal.

§9. n Choose r with Repetition.

Problems for Class. 1. $\binom{6+k-1}{k}$. Note that for 2 dice we count the 21 visual outcomes of Table 1. **2.** $\binom{26+4-1}{4} \times 1$, since there is only 1 way to put 4 given letters in alphabetical order. **3.** $10!\binom{11}{7}7!$. Since there are more men than women, first line up the 10 men in 10! ways. There are 11 spaces into which we can squeeze a woman; pick 7 of these in $\binom{11}{7}$ ways. Put the women in the selected paces in 7! ways. If we line up the women first the problem is harder and we get the answer $7!\binom{8+(10-6)-1}{10-6}10!$. **4.** $9!\binom{10}{7}7!$. **5.** After putting down the 10 consonants in 1 of the $\frac{10!}{4!2!2!}$ possible ways, we have a typical diagram such as $_\wedge N_\wedge R_\wedge C_\wedge R_\wedge R_\wedge N_\wedge C_\wedge C_\wedge L_\wedge T_\wedge$. We now have to choose 8 of the 11 spaces for the vowels and then insert the vowels. We get the solution $\frac{10!}{4!2!2!}\binom{11}{8}\frac{8!}{4!}$. **6.** $\frac{10!}{4!2!2!}\binom{11}{8}1$. **7.** $4^5 - 3[4^2]$ by considering the 3 cases: BAD__, _BAD_, and __BAD. **8.** Seating all but Paul first, we get $7!\binom{8-2}{1}1$. Other approaches give $6!\binom{7}{2}2!$ or $6!\binom{7}{1}\binom{6}{1}1$. **9.** $6!\binom{2}{1}$, or, if we glue Peter and Paul together side-by-side in 1 of the 2! possible ways, we can then treat the glued guys as 1 entity and get $2!6!$. Don't knock it, *glue* can be a very helpful aid in solving many problems. Also, using the answer to #8, we have $8! - 7!6$. **10.** Since groupings of exactly 3 or 1 of the same nationality are impossible, we have the nationalities in $2n$ pairs. (The 2 pairs of one nationality may be adjacent.) For each of the n nations, there are 2 place cards with the name of the nation and blanks for 2 names. For each of the n nations, the 4 blanks can be filled in 4! ways. The answer is $\frac{(2n)!}{2^n}(4!)^n$. **11.** By lining up the toys first, but the first 4 can be permuted, as can each next 4 in turn, we get the answer $20!/4!^5$. A longer approach that is not as elegant but gets the job done yields $\binom{20}{4}\binom{16}{4}\binom{12}{4}\binom{8}{4}\binom{4}{4}$. **12.** $18!/[5!6!7!]$. **13.** $18!/[6!6!6!3!]$. **14.** $\binom{12}{7}$. **15.** $\frac{7!}{4!2!}\binom{8}{4}1$. **16.** $\binom{5+67-1}{67}$. **17.** Suppose Alice and Eve together get k balls. This can be done in $k + 1$ ways. That leaves $30 - k$ balls for the other 2 persons, but Lucky must get at least 7 of these. So, there are $30 - k - 7$ additional balls to distribute to the Lucky and the fourth person. This can be done in $(30 - k - 7) + 1$ ways. Hence, our answer is $\sum_{k=0}^{20}(k+1)(24 - k)$. **18.** There are $\langle^3_6\rangle$ ways to pick $7 + 8 + 9 - 18$ letters from 7 A's, 8 B's, and 9 C's?

Ten Problems for Homework. 1. $\binom{26+4-1}{4}$. **2.** $\binom{26}{4}$.
3. $\binom{13}{2}\binom{4}{2}^2 44$ or $\frac{\frac{52\times3}{2}\frac{48\times3}{2}}{2}44$ or $\binom{13}{1}\binom{4}{2}\binom{12}{1}\binom{4}{2}\binom{44}{1}/2$. **4.** $5 \times 25 \times 24$, where we consider the middle letter before the first and last letters.
5. $\frac{11!}{4!4!2!} - \frac{7!}{4!2!}\binom{8}{4}$ or $\frac{7!}{4!2!}\left[\binom{8+4-1}{4} - \binom{8}{4}\right]$. **6.** $\binom{12+2-1}{2}$. **7.** $\frac{18!}{5!5!4!4!}/[2!2!]$
or $\binom{18}{5}\binom{13}{5}\binom{8}{4}\binom{4}{4}/[2!2!]$. **8.** $(mn)!/[n!^m m!]$. **9.** 6×8. **10.** One
solution is $2 + 2^2 + 2^3 + (2^4 - 2) + 2\binom{5}{2} + \binom{6}{3}$.

Five Problems for Homework. 1. Either the 2 I's are together or not. We have $\frac{6!}{4!}\langle^5_4\rangle + \frac{5!}{4!}\binom{6}{2}\langle^4_4\rangle$ from

$$\underbrace{\frac{6!}{4!}}_{\text{M,I's,PP}} \underbrace{\left(\frac{(7-2)+4-1}{4}\right)}_{\text{S's}} + \underbrace{\frac{5!}{4!}}_{\text{M,I's}} \underbrace{\binom{6}{2}}_{\text{P's}} \underbrace{\left(\frac{(8-4)+4-1}{4}\right)}_{\text{S's}}.$$

2. $\langle^6_{32}\rangle$ from $\binom{6+32-1}{32}$. **3.** $\langle^6_{26}\rangle$ from $\binom{6+26-1}{26}$. **4.** $\langle^7_{25}\rangle$ from $\binom{7+25-1}{25}$.
5. $\langle^7_{31}\rangle$ from $\binom{7+31-1}{31}$.

§10. Ice Cream Cones—The Double Dip.

1. $\langle^{12\cdot12}_5\rangle$. **2.** $\binom{12\cdot12}{5}$. **3.** $\binom{\langle^{12}_2\rangle}{5}$. **4.** $\langle^{\langle^{12}_2\rangle}_5\rangle$.

§11. Block Walking.

9. Consider the possible horizontal and vertical lines that determine the sides of the rectangles. So the answer is $\binom{11}{2}\binom{7}{2}$.
10. This is harder than #9 because width and height are not independent here. One technique is to consider where the upper left hand corner of a k–by–k square can go. We then have $(10 \times 6) + (9 \times 5) + (8 \times 4) + (7 \times 3) + (6 \times 2) + (5 \times 1)$.

§13. The Binomial Theorem.

The equality that ends the section can be proved with a block walking argument. The key is to observe that in going from $(0,0)$ to (n,n) we must pass through exactly 1 point on the line with equation $X + Y = n$, say at $(i, n - i)$. Therefore, we need to count the number of possible paths from $(0,0)$ to $(i, n - i)$ and the number of possible paths from $(i, n - i)$ to (n, n). We should find that there are $\binom{n}{i}^2$ paths from $(0,0)$ to (n, n) by way of $(i, n - i)$.

§14. Homework for a Week.

1. Think of piling the flags along with the required number of dividers to indicate where the flags go. (Our assistant will actually be putting up the flags, moving to the next pole after encountering a divider.) It is important to note that the dividers are not the flagpoles themselves but separate the flagpoles from one another. Adjacent dividers mean that a flagpole gets no flags. We have $\frac{(24+17)!}{17!}$, or, permuting the flags and then inserting the dividers, we get $24!\binom{25+17-1}{17}$. Also, put the 17 dividers down, choose 24 spaces for the flags, and then put the flags in the spaces to end up with $1\binom{18+24-1}{24}24!$. **2.** $24!\binom{23}{17}$ or $\binom{18+6-1}{6}24!$. **3.** After choosing the boxes that are to be empty, be sure first to put 1 ball into each of the remaining boxes so that none is empty: $\binom{n}{m}\binom{(n-m)+[r-(n-m)]-1}{r-(n-m)}$. **4.** $\sum_{k=0}^{10}\binom{6+(20-2k)-1}{20-2k}$ or $\sum_{r=0}^{10}\left\langle\begin{smallmatrix}6\\2r\end{smallmatrix}\right\rangle$. **5.** $\sum_{k=0}^{10}\binom{20}{k}\binom{20-k}{k}6^{20-2k}$. Note that $\binom{20}{2k}\binom{2k}{k}$ equals $\binom{20}{k}\binom{20-k}{k}$. **6.**:

$$\frac{6!}{2!2!}\left[\underbrace{\overbrace{\binom{2}{2}}_{\text{I's}}\cdot\overbrace{\binom{7-2}{1}}_{O}}_{\text{With IWI}}+\underbrace{\overbrace{\binom{2}{1}}_{I}\cdot\overbrace{\binom{7-2}{1}}_{I}\cdot\overbrace{\binom{8-3}{1}}_{O}}_{\text{Without IWI}}\right].$$

S's,N's,C,W

You might think about why the following answer is wrong: $\frac{5!}{2!2!}\times\binom{6}{3}\binom{3}{1}\times\binom{4}{1}$, arrived at by first putting down S, S, N, N, C, W together, then the I's and the O, leaving the W for last. **7.** $\left\langle\begin{smallmatrix}10\\62\end{smallmatrix}\right\rangle10^8$. **8.** $9\cdot10\cdot11-1$. **9.** $7!/4!+7!/(4!2!)$. **10.** It seems reasonable to begin with the diagram $I_\wedge I_\wedge I_\wedge I_\wedge$ and then calculate that there are $\left\langle\begin{smallmatrix}4\\4\end{smallmatrix}\right\rangle$ ways to chose the spaces for the 4 S's. So far, so good. However, it makes a difference where we put the S's in the previous step when we come to count the number of available places for the P's. For example, in the diagrams $IIIS_\wedge S_\wedge S_\wedge S_\wedge I_\wedge$ and $IS_\wedge S_\wedge S_\wedge I_\wedge S_\wedge I_\wedge I_\wedge$, we have a different number of places available for inserting P's. We abandon this disastrous attempt and start all over from the other end. We first put down the 2 P's in 1 way. From the diagram $S_\wedge P_\wedge P_\wedge$, we see that we have $\left\langle\begin{smallmatrix}3\\3\end{smallmatrix}\right\rangle$ ways to choose spaces for the 3 remaining S's after placing the first S before the 2 P's. Then from a diagram like $I_\wedge S_\wedge S_\wedge P_\wedge P_\wedge S_\wedge S_\wedge$, we see that we have $\left\langle{7\atop3}\right\rangle$ ways to place the 3 remaining I's. Note that this last count is independent of where we put the S's in the previous steps; in all cases, there are 7 places available to insert the remaining 3 I's. Finally, with 10 letters now placed, we have 11 available spaces for inserting the only M. In summary, by considering the P's, S's, I's, and the M, in turn, we get $1\left\langle{3\atop3}\right\rangle\left\langle{7\atop3}\right\rangle\left\langle{11\atop1}\right\rangle$. **11.** By putting down the P's and S's first, then selecting places for the remaining letters before putting then down, we get $1\left\langle{7\atop5}\right\rangle\left\langle{5\atop1}\right\rangle$ or doing that the other way around, we get $5\left\langle{6\atop6}\right\rangle1$. **12.** This is a Mississippi problem; think of lining up the flags and the 14 dividers. $(5+7+11+14)!/(5!7!14!)$. Another approach gives $1\cdot\left\langle{6\atop7}\right\rangle\cdot\left\langle{13\atop11}\right\rangle11!\cdot\left\langle{24\atop14}\right\rangle$. **13.** $7!\left\langle{8\atop5}\right\rangle13!$. **14.** With $0,1,2,3$ quarters, we arrive at $96+71+46+21$. **15.**:

$$\binom{5}{3}\left[\underbrace{3}_{3\cdot12+2}+\underbrace{6}_{3\cdot11+5}+\underbrace{9}_{3\cdot10+8}+\underbrace{(12-6)}_{3\cdot9+11}+\underbrace{1}_{3\cdot8+14}\right].$$

16. After picking the 2 boxes to be empty, we have 2 cases: There is a box with 3 balls or not. We might think of gluing together 3 balls or else gluing together each 2 of 2 sets of 2 balls to get

$$\binom{n}{2}\left[\binom{n}{3}(n-2)!+\frac{\binom{n}{2}\binom{n-2}{2}}{2!}(n-2)!\right].$$

Without the glue, we probably have the computation

$$\binom{n}{2}\left[\binom{n-2}{1}\binom{n}{3}\cdot(n-3)!+\binom{n-2}{2}\binom{n}{4}\binom{4}{2}\cdot(n-4)!\right].$$

17. $\left\langle{4\atop8}\right\rangle+2\left\langle{4\atop7}\right\rangle+3\left\langle{4\atop6}\right\rangle+4\left\langle{4\atop5}\right\rangle+5\left\langle{4\atop4}\right\rangle$ or $\sum_{k=0}^{4}(k+1)\binom{11-k}{3}$.
18. $4^8+\binom{8}{1}2^1 4^7+\binom{8}{2}2^2 4^6+\binom{8}{3}2^3 4^5+\binom{8}{4}2^4 4^4$ or $\sum_{j=0}^{4}\binom{8}{j}2^{16-j}$.

§15. Three Hour Exams.

Practice Hour Exam #2. 1. $11!/(4!4!2!)$. **2.** $(9+9+7+8)+1$.
3. $21!\binom{22}{5}5!$. **4.** $\binom{18}{9}$. **5.** $7!\binom{8}{6}6!$. **6.** $8!7$. **7.** $5\cdot25\cdot24\cdot23\cdot22$.
8. $\frac{11!}{4!4!2!}-\frac{7!}{4!2!}\binom{8}{4}$. **9.** $6\cdot8$. **10.** $\left\langle{7\atop32}\right\rangle$. **11.** $\left\langle{15\times15\atop7}\right\rangle$. **12.** $\left\langle{\left\langle{15\atop2}\right\rangle\atop7}\right\rangle$.
13. $\left\langle{5\atop6}\right\rangle\left\langle{5\atop9}\right\rangle$. **14.** 9. **15.** $26!\binom{25}{13}$ or $\left\langle{14\atop12}\right\rangle26!$. **16.** $\binom{12}{3}\left\langle{9\atop23}\right\rangle$. **17.** $\left\langle{10\atop62}\right\rangle10^8$.
18. $(6+8+13+14)!/(6!8!14!)$ or $\frac{(6+8+13)!}{6!8!}\left\langle{28\atop14}\right\rangle$ or $\left\langle{15\atop27}\right\rangle\frac{(6+8+13)!}{6!8!}$.

19. $26 \cdot 25^9$. **20.**:

$$\frac{9!}{4!2!} \left[\overbrace{\underbrace{1}_{\text{A's}} \cdot \underbrace{8}_{\text{U}} \cdot \underbrace{7}_{\text{E}}}^{\text{With AHA}} + \overbrace{\underbrace{2}_{\text{A}} \cdot \underbrace{8}_{\text{A}} \cdot \underbrace{8}_{\text{U}} \cdot \underbrace{7}_{\text{E}}}^{\text{Without AHA}} \right].$$

Practice Hour Exam #3. 1. $\frac{11!}{4!4!2!}$. **2.** $(9+9+7+8)+1$.
3. $21!\binom{22}{5}5!$. **4.** $7!\binom{8}{4}4!$. **5.** $7!\binom{8}{6}6!$. **6.** $8!7$ or $7!\binom{8}{2}2!$ or $9! - 8!2!$.
7. $26 \cdot 25 \cdot 24 \cdot 24 \cdot 24$. **8.** $\frac{7!}{3!4!}\left\langle\begin{smallmatrix}9\\3\end{smallmatrix}\right\rangle\binom{3}{1}$ or $\left\langle\begin{smallmatrix}4\\4\end{smallmatrix}\right\rangle\left\langle\begin{smallmatrix}9\\3\end{smallmatrix}\right\rangle\binom{3}{1}$. **9.** $6 \cdot 8$. **10.** $\binom{3}{1}5!5!$
or $\binom{5}{4}4! \cdot \binom{5}{3}3! \cdot 3!$. **11.** $\left\langle\begin{smallmatrix}9\\2\end{smallmatrix}\right\rangle\left\langle\begin{smallmatrix}\\7\end{smallmatrix}\right\rangle$. **12.** $\left\langle\begin{smallmatrix}3\\4\end{smallmatrix}\right\rangle4!$ or $6!/2$. **13.** $1 \cdot \binom{19}{5}$.
14. $\binom{25}{6} - \binom{20}{6}$. **15.** $\binom{20}{6}$. **16.** $\frac{(4+6+7)!}{4!6!7!}\binom{18}{5}$. **17.** $\frac{(4+6+7)!}{4!6!7!}\left\langle\begin{smallmatrix}14\\5\end{smallmatrix}\right\rangle$.
18. $2 \cdot 7!7!$. **19.** $6!7!$. **20.** $2^7 6!$. **21.** $\binom{5}{2}\binom{6}{2}$. **22.** Pick the n top flags
for the n poles in $\binom{m}{n}$ ways. We have $m - n$ remaining
distinguishable flags to put on the n flagpoles, which are now
distinguishable by their top flags. Altogether, we have
$\binom{m}{n}\frac{[(m-n)+(n-1)]!}{(n-1)!}$. **23.** $\frac{(8+9+10+10)!}{8!9!10!}$ or, for comparison with the next
2 questions, $\frac{(8+9+10)!}{8!9!}\left\langle\begin{smallmatrix}28\\10\end{smallmatrix}\right\rangle$ or $\left\langle\begin{smallmatrix}11\\27\end{smallmatrix}\right\rangle\frac{(8+9+10)!}{8!9!}$. **24.** $\frac{(8+9+10)!}{8!9!}\binom{26}{10}$ or
$\left\langle\begin{smallmatrix}11\\16\end{smallmatrix}\right\rangle\frac{(8+9+10)!}{8!9!}$. **25.** $\left\langle\begin{smallmatrix}11\\5\end{smallmatrix}\right\rangle\frac{(8+9+10)!}{8!9!}$.

Chapter 2. Principle of Inclusion and Exclusion

§16. Introduction to PIE.

PIE Problems I. 1. $10^9 - 10 \cdot 9!$. **2.** $5 \cdot 26^3 + 5 \cdot 26^3 - 5^2 26^2$.
3. $21^2 26^2$. We can also subtract the answer to #2 from 26^4. **4.** Here
the universal set is all the placements of the 25 red balls into 3
distinguishable boxes without restriction. We want to count those
placements in none of the sets of all placements where a particular
box gets more than 15 balls. (Set A_i is the set of placements where
box i is "overloaded," that is, has at least 16 balls.) Our answer is
$\left\langle\begin{smallmatrix}3\\25\end{smallmatrix}\right\rangle - \binom{3}{1}\left\langle\begin{smallmatrix}3\\9\end{smallmatrix}\right\rangle$. Be sure that you can explain the "$\left\langle\begin{smallmatrix}3\\9\end{smallmatrix}\right\rangle$" in the
solution; the most common mistake here is to have "$\left\langle\begin{smallmatrix}2\\9\end{smallmatrix}\right\rangle$" in place of
"$\left\langle\begin{smallmatrix}3\\9\end{smallmatrix}\right\rangle$". **5.** $\left\langle\begin{smallmatrix}3\\25\end{smallmatrix}\right\rangle - \binom{3}{1}\left\langle\begin{smallmatrix}3\\14\end{smallmatrix}\right\rangle + \binom{3}{2}\left\langle\begin{smallmatrix}3\\3\end{smallmatrix}\right\rangle$. **6.** $\left\langle\begin{smallmatrix}5\\9\end{smallmatrix}\right\rangle\frac{9!}{2!} + \left\langle\begin{smallmatrix}5\\9\end{smallmatrix}\right\rangle\frac{9!}{2!} - \binom{4}{2} \cdot 1 \cdot \left\langle\begin{smallmatrix}7\\7\end{smallmatrix}\right\rangle7!$ or
$\frac{9!}{2!}\left\langle\begin{smallmatrix}10\\4\end{smallmatrix}\right\rangle + \frac{9!}{2!}\left\langle\begin{smallmatrix}10\\4\end{smallmatrix}\right\rangle - 7!\binom{8}{6}\binom{4}{2}1$. **7.** $\left\langle\begin{smallmatrix}3\\18\end{smallmatrix}\right\rangle - \binom{3}{1}\left\langle\begin{smallmatrix}3\\8\end{smallmatrix}\right\rangle$. **8.** $8! + 8! - 6!$.
9. $[3 \cdot 8!] - [6! + 6! + 6!] + [4!]$. **10.** $6^7 - 5^4 \cdot 6 \cdot 6^2$.

§17. Proof of PIE.

PIE Problems II. 1. For each letter Z, let A_Z be the set of
10-letter words without a Z. We want $|A_A \cup A_E \cup A_I \cup A_O \cup A_U|$. So

we get $\binom{5}{1}25^{10} - \binom{5}{2}24^{10} + \binom{5}{3}23^{10} - \binom{5}{4}22^{10} + \binom{5}{5}21^{10}$ or
$\sum_{k=1}^{5}(-1)^{k+1}\binom{5}{k}(26-k)^{10}$. **2.** Let A_i be unrestricted placements such that box i gets more that 15 balls. We count the placements that are in none of the A_i to get $\left\langle{4\atop40}\right\rangle - \binom{4}{1}\left\langle{4\atop24}\right\rangle + \binom{4}{2}\left\langle{4\atop8}\right\rangle$. **3.** Let A_i consist of the unrestricted placements such that box i remains empty. We want $|A_1 \cup A_2 \cup A_3 \cup A_4 \cup A_5 \cup A_6|$, and so we get $\binom{6}{1}5^{11} - \binom{6}{2}4^{11} + \binom{6}{3}3^{11} - \binom{6}{4}2^{11} + \binom{6}{5}1^{11}$ or $\sum_{k=1}^{5}(-1)^{k+1}\binom{6}{k}(6-k)^{11}$. **4.** For each j, let A_j be the solutions we get by ignoring all the restrictions $x_i \le 30$ such that $x_j > 30$. We count the solutions that are in none of the A_j to get $\left\langle{4\atop100}\right\rangle - \binom{4}{1}\left\langle{4\atop69}\right\rangle + \binom{4}{2}\left\langle{4\atop38}\right\rangle - \binom{4}{3}\left\langle{4\atop7}\right\rangle$ or $\sum_{k=0}^{3}(-1)^{k}\binom{4}{k}\left\langle{4\atop100-31k}\right\rangle$.
5. $\left\langle{3\atop10}\right\rangle - \left[\left\langle{3\atop2}\right\rangle + \left\langle{3\atop1}\right\rangle + \left\langle{3\atop2}\right\rangle\right] + 0$. Think of an unlimited supply (actually, 10 of each will do) of each type of coin and let A_{p}, for example, be the selections of 10 coins that have at least 8 pennies. The number of these selections is $\left\langle{3\atop2}\right\rangle$ and is the first term within the square brackets. **6.** Let A_i be the set of unrestricted distributions that have box i empty. We count the distributions that are in none of the A_i to get $\sum_{k=0}^{n}(-1)^{k}\binom{n}{k}(n-k)^{r}$.

§18. Derangements.

PIE Problems III. 1. $\left\langle{4\atop20}\right\rangle - \left[2\cdot\left\langle{4\atop7}\right\rangle + 2\cdot\left\langle{4\atop11}\right\rangle\right] + \left\langle{4\atop2}\right\rangle$. **2.** We use PIE with A_i denoting the 12-term sequences of digits that are missing digit i. We want to count the elements in $A_0 \cup A_1 \cup \cdots \cup A_9$. We get $\sum_{k=1}^{9}(-1)^{k+1}\binom{10}{k}(10-k)^{21}$.
3. $5! - \binom{3}{1}2\cdot4! + \binom{3}{2}2^2\cdot3! - \binom{3}{3}2^3\cdot2!$. **4.** $3\cdot\left\langle{5\atop9}\right\rangle\frac{9!}{2!} - \binom{3}{2}\cdot\left\langle{7\atop7}\right\rangle7! + 0$ or $3\cdot\frac{9!}{2!}\left\langle{10\atop4}\right\rangle1 - 3\cdot7!\left\langle{8\atop6}\right\rangle1 + 0$. **5.** See Figure 1; also, $\left[\frac{9!}{2!}\left\langle{10\atop4}\right\rangle + \frac{9!}{2!}\left\langle{10\atop4}\right\rangle + \frac{10!}{2!2!}\left\langle{11\atop3}\right\rangle\right] - \left[7!\left\langle{8\atop6}\right\rangle + \frac{8!}{2!}\left\langle{9\atop5}\right\rangle + 6!\left\langle{7\atop4}\right\rangle\left\langle{11\atop3}\right\rangle\right] + \left[6!\left\langle{7\atop7}\right\rangle\right]$.

PIE Problems IV. 1. We line up the 8 integers and consider dropping some glue in the spaces between some of the integers. We look at the diagram $1^{\vee}2^{\vee}3^{\vee}4^{\vee}5^{\vee}6^{\vee}7^{\vee}8$ and let A_i be the set of permutations in which i is immediately followed by $i+1$ for $i = 1,2,3,4,5,6,7$, which corresponds to having glue in the ith wedge. (The wedges are inverted to our usual position only because our glue does not run uphill.) The 7 A_i are the sets of permutations that contain, respectively, the 7 strings 12, 23, 34, 45, 56, 67, 78. Each time we put glue in a new spot we have a net loss of 1 symbol; with glue in k spots, we are permuting $8 - k$ symbols. We want to count the permutations that can be achieved without any

$$\left[\underbrace{\left\langle{5\atop9}\right\rangle\frac{9!}{2!}}_{\text{C,C,O,O}} + \underbrace{\left\langle{5\atop9}\right\rangle\frac{9!}{2!}}_{\text{O,O,I,I}} + \underbrace{\left\langle{4\atop10}\right\rangle\frac{10!}{2!2!}}_{\text{I,I,S}} \right]$$

$$-\left[\underbrace{\left\langle{7\atop7}\right\rangle 7!}_{\text{C,C,O,O,I,I}} + \underbrace{\left\langle{6\atop8}\right\rangle\frac{8!}{2!}}_{\text{O,O,I,I,S}} + \underbrace{\overbrace{1}^{\text{C,C,O,O}}\cdot\left\langle{5\atop3}\right\rangle\overbrace{1}^{\text{I,I,S}}\cdot\overbrace{\left\langle{8\atop6}\right\rangle 6!}^{\text{the rest}}}_{\text{C,C,O,O \& I,I,S}} \right]$$

$$+\left[\underbrace{\left\langle{8\atop6}\right\rangle 6!}_{\text{C,C,O,O,I,I,S}} \right]$$

FIGURE 1.

glue, that is, permutations in none of the A_i. So, by the Lemma for PIE, we obtain $\sum_{k=0}^{7}(-1)^k\binom{7}{k}(8-k)!$. **2.** Now there are 8 glue spots. This time we get $\left[\sum_{k=0}^{7}(-1)^k\binom{8}{k}(7-k)!\right]+1$.

3. $\sum_{k=0}^{22}(-1)^k\binom{23}{k}(23-k)^{37}$ by thinking of the A_i being the unrestricted distributions where boy i gets no book. **4.** We change the form of the problem in order to simplify it. Let $x_1 = 1 + y_1$; so $0 \le y_1 \le 4$. Let $x_2 = 2 + y_2$; so $0 \le y_2 \le 5$. Let $x_3 = 3 + y_3$; so $0 \le y_3 \le 6$. Let $x_4 = 5 + y_4$; so $0 \le y_4$. Thus, we are looking at solutions for the system

$$y_1 + y_2 + y_3 + y_4 = 29,$$
$$0 \le y_1 \le 4, \quad 0 \le y_2 \le 5, \quad 0 \le y_3 \le 6, \quad 0 \le y_4.$$

So, $\left\langle{4\atop29}\right\rangle - \left[\left\langle{4\atop24}\right\rangle + \left\langle{4\atop23}\right\rangle + \left\langle{4\atop22}\right\rangle\right] + \left[\left\langle{4\atop18}\right\rangle + \left\langle{4\atop17}\right\rangle + \left\langle{4\atop16}\right\rangle\right] - \left\langle{4\atop11}\right\rangle$ is the desired number. Here, the $\left\langle{4\atop23}\right\rangle$ comes from first putting 6 1's in the y_2 box and then putting, with repetition allowed, the remaining 23 1's into the 4 boxes. Note that it is $\left\langle{4\atop23}\right\rangle$ and not $\left\langle{3\atop23}\right\rangle$ because more than 6 1's can go into the y_2 box altogether. **5.** We consider the cases with no box empty, with exactly 1 box empty, and with 2 boxes empty, in turn, to get $\frac{3^{10}-3\cdot2^{10}+3}{3!} + \frac{2^{10}-2}{2!} + 1$. **6.** $1 - (1/e)$ is about 63%.

7. $D_n = nD_{n-1} + (-1)^n = (n-1)D_{n-1} + D_{n-1} + (-1)^n = (n-1)D_{n-1} + [(n-1)D_{n-2} + (-1)^{n-1}] + (-1)^n = (n-1)[D_{n-1} + D_{n-2}].$

§21. A Plethora of Problems.

1. The question is, How many ways are there to put m distinguishable balls into n distinguishable boxes with no box empty? Thinking of k as the number of excluded boxes, we have the answer $\sum_{k=0}^{n} (-1)^k \binom{n}{k} (n-k)^m$. **2.** Considering the diagram $_\wedge T _\wedge T _\wedge T _\wedge T _\wedge T _\wedge T _\wedge T _\wedge$, we want to put 17 indistinguishable balls (H's) into 9 distinguishable boxes with the restriction of at most 4 balls in each box. Without the restriction, suppose at least k particular boxes get at least 5 balls. We get $\sum_{k=0}^{3} (-1)^k \binom{9}{k} \left\langle \begin{smallmatrix} 9 \\ 17 \ 5k \end{smallmatrix} \right\rangle$.
3. Thinking of at least some particular k of the given 6 numbers in their natural position, we use the Lemma for PIE to get $\sum_{k=0}^{6} (-1)^k \binom{6}{k} (12-k)!$. **4.** $\sum_{k=0}^{12} (-1)^k \binom{12}{k} (2^k)(24-k-1)!$
5. Supposing that at least k particular pairs are maintained, we get $\sum_{k=0}^{5} (-1)^k \binom{5}{k} \frac{(10-2k)!}{2^{5-k}(5-k)!}$. **6.** We pass out 5 right gloves in 6! ways, keeping 1 for ourself. In passing out the left gloves to the 5 persons and ourself, either we get a matching pair or not. The solution is $6![D_5 + D_6]$. **7.** Thinking of at least k particular persons getting a matching pair, we have $\sum_{k=0}^{5} (-1)^k \binom{5}{k} \cdot \binom{6}{k} k! \cdot \frac{(12-2k)!}{2^{!6-k}}$. **8.** We can consider glue in at least k particular spots of the 8 spots in the diagram $0^\vee 1^\vee 2^\vee 3^\vee 4^\vee 5^\vee 6^\vee 7^\vee 8$ but remember that each arrangement must begin with 0. We have $\sum_{k=0}^{8} (-1)^k \binom{8}{k} (8-k)!$.
9. $2\left[\frac{5!}{3!2!} \left\langle \begin{smallmatrix} 7 \\ 12 \end{smallmatrix} \right\rangle \frac{12!}{4!2!}\right] - \left[\frac{5!}{3!2!} \cdot \left\langle \begin{smallmatrix} 7 \\ 6 \end{smallmatrix} \right\rangle \frac{5!}{3!2!} \cdot \left\langle \begin{smallmatrix} 13 \\ 6 \end{smallmatrix} \right\rangle 6!\right]$.
10. $2\left[\frac{5!}{3!2!} \left\langle \begin{smallmatrix} 7 \\ 12 \end{smallmatrix} \right\rangle \frac{12!}{4!2!}\right] - 2\left[\frac{5!}{3!2!} \cdot \left\langle \begin{smallmatrix} 7 \\ 6 \end{smallmatrix} \right\rangle \frac{5!}{3!2!} \cdot \left\langle \begin{smallmatrix} 13 \\ 6 \end{smallmatrix} \right\rangle 6!\right]$.
11. Subtract the answer to #9 from $\frac{18!}{4!4!2!2!}$. **12.** The 2 cases of first R before first E and first E before first R are symmetric. So, putting the R's, E's, and C's down first, we calculate $2\left[\frac{5!}{3!2!} \left\langle \begin{smallmatrix} 7 \\ 3 \end{smallmatrix} \right\rangle\right] \cdot \left\langle \begin{smallmatrix} 11 \\ 8 \end{smallmatrix} \right\rangle \frac{8!}{2!}$. **13.** $\frac{9!}{3!2!4!} \cdot \left\langle \begin{smallmatrix} 11 \\ 8 \end{smallmatrix} \right\rangle \frac{8!}{2!}$. **14.** $1 \cdot \left\langle \begin{smallmatrix} 13 \\ 6 \end{smallmatrix} \right\rangle 6!$.
15. $\left[1 \cdot \left\langle \begin{smallmatrix} 9 \\ 10 \end{smallmatrix} \right\rangle \frac{10!}{2!2!} + 1 \cdot \left\langle \begin{smallmatrix} 7 \\ 12 \end{smallmatrix} \right\rangle \frac{12!}{4!2!} + 1 \cdot \left\langle \begin{smallmatrix} 5 \\ 14 \end{smallmatrix} \right\rangle \frac{14!}{4!4!}\right] - \left[1\left\langle \begin{smallmatrix} 9 \\ 10 \end{smallmatrix} \right\rangle \frac{10!}{4!} + 1\left\langle \begin{smallmatrix} 9 \\ 4 \end{smallmatrix} \right\rangle 1 \cdot \left\langle \begin{smallmatrix} 13 \\ 6 \end{smallmatrix} \right\rangle 6! + 1 \cdot \left\langle \begin{smallmatrix} 11 \\ 8 \end{smallmatrix} \right\rangle \frac{8!}{2!}\right] + \left[1\left\langle \begin{smallmatrix} 13 \\ 6 \end{smallmatrix} \right\rangle 6!\right]$.
16. Subtract the answer to #15 from $\frac{18!}{4!4!2!2!}$.
17. The 3 terms in the first factor are motivated by first putting down the N's, C's, and E's before putting in the R's in the 3 diagrams $_\wedge E \ldots$ and $_\wedge N _\wedge E \ldots$ and $_\wedge N _\wedge N _\wedge E \ldots$, respectively. In each diagram, at least 1 R must go into at least 1 of the spaces indicated by the wedges. We get

$\left\{\frac{3!}{2!}\left\langle\frac{5}{3}\right\rangle\left[\left\langle\frac{9}{4}\right\rangle-\left\langle\frac{8}{4}\right\rangle\right]+\frac{6!}{3!2!}\left[\left\langle\frac{9}{4}\right\rangle-\left\langle\frac{7}{4}\right\rangle\right]+\frac{5!}{3!2!}\left[\left\langle\frac{9}{4}\right\rangle-\left\langle\frac{6}{4}\right\rangle\right]\right\}\cdot\left\langle\frac{13}{6}\right\rangle 6!.$

18. $\left\{\frac{7!}{3!4!}\left\langle\frac{9}{10}\right\rangle\frac{10!}{2!2!}+\frac{5!}{3!2!}\left\langle\frac{7}{12}\right\rangle\frac{12!}{4!2!}+\frac{3!}{2!}\left\langle\frac{5}{14}\right\rangle\frac{14!}{4!4!}\right\}-$

$\left\{\left[\frac{3!}{2!}\left\langle\frac{5}{3}\right\rangle+\frac{5!}{3!2!}\left\langle\frac{7}{1}\right\rangle\right]\left\langle\frac{9}{10}\right\rangle\frac{10!}{4!}+\frac{7!}{3!4!}\left\langle\frac{9}{4}\right\rangle\frac{3!}{2!}\left\langle\frac{13}{6}\right\rangle 6!+\frac{5!}{3!2!}\left\langle\frac{7}{3}\right\rangle\left\langle\frac{11}{8}\right\rangle\frac{8!}{2!}\right\}+$

$\left\{\frac{3!}{2!}\left\langle\frac{5}{3}\right\rangle\cdot\left\langle\frac{9}{3}\right\rangle+\frac{6!}{3!2!}\left[\left\langle\frac{9}{4}\right\rangle-\left\langle\frac{7}{4}\right\rangle\right]+\frac{5!}{3!2!}\left[\left\langle\frac{9}{4}\right\rangle-\left\langle\frac{6}{4}\right\rangle\right]\right\}\left\langle\frac{13}{6}\right\rangle 6!.$

§22. Eating Out.

Related Problem. For a subset of q of the $2n$ conditions, how many of all $n!$ permutations satisfy these q conditions? The answer to this question is either $(n-q)!$ or 0, depending on whether the q conditions are compatible or not. However, we notice that it is exactly the pairs of consecutive conditions that are incompatible, taking "1 is 1st" to follow "n is 1st" in cyclic order. We have encountered this situation before. This is analogous to choosing q of $2n$ knights sitting at a round table without selecting any pair of adjacent knights. See §12. From page 21, we know that this can be done in $\frac{2n}{2n-q}\binom{2n-q}{q}$ ways. Hence, the solution to our related problem is $\sum_{q=0}^{n}(-1)^q\left[\frac{2n}{2n-q}\binom{2n-q}{q}\right](n-q)!$, or

$$2n\sum_{q=0}^{n}(-1)^q\frac{(n-q)!}{2n-q}\binom{2n-q}{q}.$$

Perhaps we can see why this answer can be obtained simply by dividing the solution to the Ménage Problem by $(n-1)!$. Consider the diagram

$$\cdots \underset{4}{\wedge} W_3 \underset{3}{\wedge} W_2 \underset{2}{\wedge} W_1 \underset{1}{\wedge} W_n \underset{n}{\wedge} W_{n-1} \underset{n-1}{\wedge} \cdots$$

formed into a loop as representing the situation in placing the persons in the Ménage Problem, where the n numbered wives have already been placed at the round table in $(n-1)!$ ways and there remains only the placing of the n correspondingly numbered husbands into the numbered available spaces. The forbidden placements of the husbands are the $2n$ conditions of the Related Problem. It follows that the solution to the Ménage Problem is $(n-1)!$ times the solution to the Related Problem.

Eating Out Problems.

1a. $\sum_{k=0}^{n-1}(-1)^k\binom{n-1}{k}(n-k)!=D_n+D_{n-1}.$

1b. Let the solution $\left[\sum_{k=0}^{n-1}(-1)^k\binom{n}{k}(n-k-1)!\right]+(-1)^n$ be denoted by A_n and show that $A_n + A_{n-1} = D_{n-1}$.

1c. $\left[\sum_{k=0}^{n-1}(-1)^k\binom{n}{k}(n-k)!\right]+0 = D_n - (-1)^n = nD_{n-1}$.

1d. $\sum_{k=0}^{n-1}(-1)^k\binom{n-1}{k}(n-k-1)! = D_{n-1}$.

2a. Select k pairs, pair the rest, and seat all in $\binom{n}{k}\cdot\frac{(2n-2k)!}{2!^{n-k}(n-k)!}\cdot(n-1)!2!^{n-1}$ ways. The solution is $\frac{1}{2n}\sum_{k=0}^n(-2)^k\binom{n}{k}^2 k!(2n-2k)!$, and, if we assume $\Pi_{r=1}^0(2r-1)=1$, then our solution becomes $(n-1)!2^{n-1}\sum_{k=0}^n\left[(-1)^k\binom{n}{k}\Pi_{r=1}^{n-k}(2r-1)\right]$.

2b. $\sum_{k=0}^n(-1)^k\binom{n}{k}(2n-k-1)!$. **2c.** $\sum_{k=0}^n(-1)^k\binom{n}{k}2^k(2n-k-1)!$.

2d. Seating men first, then women, we get $(n-1)!D_n$.

2e. $\sum_{k=0}^n(-1)^k\binom{n}{k}(2n-k)!$. **2f.** $\sum_{k=0}^n(-1)^k\binom{n}{k}2^k(2n-k)!$.

2g. With the diagram $\underline{H}W\underline{H}W \ldots \underline{H}W\underline{H}W$ and the diagram $W\underline{H}W\underline{H} \ldots W\underline{H}W\underline{H}$ to help answer the where-question, we have the direct calculation

$$\left\{\sum_{k=0}^{n-1}(-1)^k\underbrace{\binom{n}{k}}_{\text{who}}\underbrace{\left[\binom{n}{k}+\binom{n-1}{k}\right]}_{\text{where}}\underbrace{k!(n-k)!^2}_{\text{how}}\right\}+(-1)^n n!.$$

A different approach begins by considering the 3 possible diagrams: first $H\ldots W$, second $W_i\ldots H_j$ with $i\neq j$, and third $W_i\ldots H_i$. For each of the $n!$ permutations of the n husbands that fill in the diagrams, we then have the following number of ways to fill in the wives. For the first diagram, we have D_n ways. For the second diagram we have D_n ways again because there is a 1·1 correspondence with the first diagram (move the H_j to the front of W_i). For the third diagram, we have D_{n-1} ways, as follows from the first diagram after ignoring the W_i and H_i. So, our solution can be simplified to the expression $n![2D_n + D_{n-1}]$.

Setting our 2 solutions equal to each other may be the easiest way to prove the equation stated in #1a above.

2h. Analogous to the solution of the Ménage Problem—but here the "2" comes from the 2 cases of whether a wife or a husband is in the first seat—we calculate $S_k = \binom{n}{k}\left[2(n-k)!^2\langle\binom{2(n-k)+1}{k}\rangle k!\right]$ to get the solution $2n!\sum_{k=0}^n(-1)^k\binom{2n-k}{k}(n-k)!$.

2i. For a given permutation $H_1H_2\ldots H_n$ of the n husbands, where, in turn, can wives W_1, W_2, $\ldots$, and W_n be placed? We get $[n!][1\cdot 3\cdot 5\cdots(2n-1)]$, or $(2n)!/2^n$.

2j. For a given permissible arrangement $\_H_1\_H_2\_H_3\ldots\_H_n$ of the n husbands, where, in turn, can wives W_1, W_2, $\ldots$ and W_n be placed? Our solution is $n!$.

Chapter 3. Generating Functions

§26. Clotheslines.

Homework. The relevant question is, What is the generating function for the sequence $\{a_r\}$? **1.** Coefficient of z^r in $(z^3+z^4+z^5)^6$ or in $z^{18}(1+z+z^2)^6$. **2.** Coefficient of z^r in $(z^0+z^1+z^2+z^3+\cdots)^8$ or in $\left[\frac{1}{1-z}\right]^8$. Of course, the coefficient is $\left\langle{8\atop r}\right\rangle$. **3.** The generating function for the first 2 boxes together is $(z^0+z^1+z^2)^2$. The solution is the coefficient of z^r in $(z^0+z^1+z^2)^2(z^0+z^1+z^2+z^3+\cdots)^{n-2}$ or in $(1+z+z^2)^2\left[\frac{1}{1-z}\right]^{n-2}$. **4.** Coefficient of z^r in

$$[\underbrace{z^0}_{0,0}+(\underbrace{z^1}_{1,0}+\underbrace{z^1}_{0,1})+(\underbrace{z^2}_{2,0}+\underbrace{z^2}_{1,1}+\underbrace{z^2}_{0,2})][z^0+z^1+z^2+z^3+\cdots]^{n-2},$$

or $[1+2z+3z^2]\left[\frac{1}{1-z}\right]^{n-2}$. **5.** Coefficient of z^{10} in $g(z)$ where $g(z)=[z^0+z^1+z^2+\cdots][z^1+z^2+z^3+\cdots][z^0+z^1][z^0+z^2][z^0+z^2+z^4+z^6+\cdots]$

$$=\frac{1}{1-z}\cdot\frac{z}{1-z}\cdot(1+z)\cdot(1+z^2)\cdot\frac{1}{1-z^2}$$

$$=(z+z^3)\left(\frac{1}{1-z}\right)^3=(z+z^3)\sum_{k=0}^{\infty}\left\langle{3\atop k}\right\rangle z^k.$$

The desired coefficient turns out to be $\left\langle{3\atop 9}\right\rangle+\left\langle{3\atop 7}\right\rangle$.

§27. Examples and Homework.

Homework. 1. Coefficient of z^{10} in $g(z)$ where

$$g(z)=\frac{1-z^8}{1-z}\cdot\frac{1-z^9}{1-z}\cdot\frac{1-z^8}{1-z}$$

$$=\frac{(1-z^8)^2(1-z^9)}{(1-z)^3}.$$

2. Coefficient of z^r in $\left(\frac{1}{1-z}\right)^3$. Note that the question concerns r coins and not r cents. **3.** Coefficient of z^r in $(1 + z^5 + z^9)^{10}$.
4. $g(z) = \frac{1}{1-z^2} \cdot \frac{1}{1-z^3} \cdot \frac{1}{1-z^4}$. **5.** For Lucky Pierre, and for each of the other 3 persons as well, the generating function for selecting balls is within the square brackets of the following, with the exponent 4 accounting for Lucky and the other 3 persons.

$$\left[\underbrace{(z^2 + z^3 + z^4 + \cdots)}_{\text{Lucky's red balls}} \underbrace{(z^2 + z^3 + z^4 + \cdots)}_{\text{Lucky's blue balls}}\right]^4 .$$

We want the coefficient of z^n in the generating function $\frac{z^{16}}{(1-z)^8}$. Rather than first focusing on Lucky, we might first focus on distributing the red balls and then the blue balls. From this approach we get the generating function $\left(\frac{z^2}{1-z}\right)^4 \left(\frac{z^2}{1-z}\right)^4$. The form $z^{16} \left(\frac{1}{1-z}\right)^8$ of the generating function tells us that the desired coefficient is $\left\langle{8 \atop n-16}\right\rangle$. Now, if not before, we see from the answer that we should have been able to answer the given question immediately.
6. $g(z) = \left[\frac{1}{1-z} \cdot \frac{1}{1-z}\right]^5 = \left(\frac{1}{1-z}\right)^{10}$. The desired coefficient is $\left\langle{10 \atop n}\right\rangle$. Can you give an elementary explanation for this answer?
7. $g(z) = (z^1 + z^2 + z^3 + \cdots + z^6)^{10} = z^{10}(1 + z + z^2 + \cdots + z^5)^{10} = z^{10}\left(\frac{1-z^6}{1-z}\right)^{10} = z^{10}(1-z^6)^{10}\left(\frac{1}{1-z}\right)^{10}$. **8.** Guided by the hint, we want to count the number of nonnegative integer solutions to $x_1 + x_2 + x_3 + x_4 + x_5 + x_6 = r$, where $r = 10$, under the condition that $1x_1 + 2x_2 + 3x_3 + 4x_4 + 5x_5 + 6x_6$ is even. Since each of $2x_2$, $4x_4$, and $6x_6$ is obviously even, the condition reduces to $x_1 + 3x_3 + 5x_5$ is even. Hence, either exactly 1 of x_1, x_3, and x_5 is even or else all 3 are even. We want the coefficient of z^{10} in

$$\left(\frac{1}{1-z}\right)^3 \left[\binom{3}{1}\left(\frac{1}{1-z^2}\right)\left(\frac{z}{1-z^2}\right)^2 + \left(\frac{1}{1-z^2}\right)^3\right].$$

More Homework. 1. The coefficient of z^r in $g(z)$ where $g(z) = \frac{z^2}{1-z^2} \cdot \frac{z^3}{1-z^3} \cdot \frac{z^4}{1-z^4}$ is the number of positive integer solutions to the equation $2x_1 + 3x_2 + 4x_3 = r$. We want the coefficient of z^{66} in $g(z)$. **2.** Coefficient of z^{40} in $(z^1 + z^2 + \cdots + z^6)^{10}$. **3.** Coefficient

of z^{20} in $[1 + 2z + 3z^2 + 4z^3 + 5z^4 + 6z^5][\binom{2}{1} \cdot \frac{1}{1-z^2} \frac{z}{1-z^2}]$.
4. Coefficient of z^{25} in $(3z^2 + z^3 + 2z^4)^7$. **5.** Coefficient of z^{20} in
$\left[\sum_{k=0}^{5} \langle^3_k\rangle z^k\right] \left[\sum_{k=4}^{\infty}(k+1)z^k\right]$. **6.** Why would these dice change a
game of Monopoly? **7.** We want the coefficient of b^2w^2 in $(b+w)^4$,
or we want the coefficient of z^2 in $(1+z)^4$.

§29. Exponential Generating Functions.

Three Exercises. 1. Coefficient of $z^r/r!$ in egf
$(e^z - 1) \cdot \frac{e^z + e^{-z}}{2} \cdot \frac{e^z - e^{-z}}{2} \cdot \left(1 + \frac{z^2}{2!}\right)^2$. **2.** Coefficient of $z^r/r!$ in
$(e^z - z)^5$. **3.** Coefficient of $z^r/r!$ in egf $\frac{e^z + e^{-z}}{2}(e^z - 1)e^{2z}$.

Four Exercises. 1. Coefficient of $z^r/r!$ in egf $\left(\frac{z^2}{2!} + \frac{z^3}{3!} + \frac{z^4}{4!}\right)^6$.
2. Coefficient of $z^r/r!$ in egf $e^{6z} - (e^z - 1)^6$. **3.** Coefficient of $z^r/r!$
in egf $e^{4z}\left(\frac{e^z + e^{-z}}{2} - 1\right)$, which is the coefficient of $z^r/r!$ in
$(e^{5z} + e^{3z} - 2e^{4z})/2$, which is $(5^r + 3^r - 2^{2r+1})/2$. **4.** Coefficient of
$z^{25}/25!$ in cgf
$(e^z - 1 - z)\left(1 + z + \frac{z^2}{2!} + \frac{z^3}{3!}\right)\frac{e^z - e^{-z}}{2}e^z\left(\frac{z^7}{7!} + \frac{z^8}{8!} + \frac{z^9}{9!} + \frac{z^{10}}{10!} + \frac{z^{11}}{11!}\right)$.

Five Review Exercises. 1. Coefficient of $z^5/5!$ in
$(1 + z + \frac{z^2}{2} + \frac{z^3}{6})^2(1 + z + \frac{z^2}{2})^2(1+z)^2$. **2.** Coefficient of $z^{34}/34!$ in
$(1 + z + \frac{z^2}{2!} + \frac{z^3}{3!} + \frac{z^4}{4!})(e^{4z})(\frac{z^8}{8!})$, or coefficient of $z^{26}/26!$ in
$\binom{34}{12}(1 + z + \frac{z^2}{2!} + \frac{z^3}{3!} + \frac{z^4}{4!})e^{4z}$. **3.** Coefficient of z^{35} in
$\left[\sum_{k=7}^{\infty}(k+1)z^k\right]\left[\frac{z^4}{1-z}\right]\left[\frac{1}{1-z}\right]^3$. **4.** Coefficient of $z^{25}/25!$ in
$\left(\frac{e^z - e^{-z}}{2}\right)^5$. **5.** Coefficient of z^{10} in
$(z^3 + z^4 + 2z^5 + 2z^6 + 2z^7 + z^8 + z^9)\frac{z^3}{1-z^3}\left(\frac{z}{1-z}\right)^2$.

§30. Comprehensive Exams.

A Warm-up Exercise. 1. 3^4. **2.** D_4. **3.** $\binom{14}{5}3^5 5^9$. **4.** $\langle^3_5\rangle\langle^5_9\rangle$.
5. $\langle^6_5\rangle\langle^{10}_9\rangle$.

Core Exam #1. 1. $5 \cdot 7 + 5 \cdot 4 + 7 \cdot 4$, or $\binom{16}{2} - [\binom{5}{2} + \binom{7}{2} + \binom{4}{2}]$.
2. $6 \cdot 8 - 1$. **3.** $8!\binom{9}{5}5!$. **4.** $\langle^{26}_5\rangle$. **5.** $18!/(5!5!2!4!4!2!)$.
6. $\frac{11!}{4!4!2!} - \frac{7!}{4!2!}\binom{8}{4}$, or $\frac{7!}{4!2!}[\langle^8_4\rangle - \binom{8}{4}]$. **7.** $20!$. **8.** $1 - \langle^{10}_4\rangle \cdot \frac{4!}{2!}$.
9. $9! - 2 \cdot 8!$, or $8!7$. **10.** $26!/16!$. **11.** $\frac{(7+20+9)!}{7!20!9!}$. **12.** $\frac{(7+20+12)!}{7!20}\binom{38}{9}$.
13. $\langle^3_{14}\rangle$. **14.** $\langle^6_9\rangle$. **15.** $\langle^9_{12}\rangle9^{13}$. **16.** $10!$. **17.** $\langle^5_{12}\rangle^4$. **18.** $\langle^9_8\rangle$. **19.** $\binom{16}{5}$.

20. $\left[\binom{5}{9}\frac{9!}{2!} + \binom{4}{10}\frac{10!}{2!2!}\right] - \binom{5}{3}\binom{8}{6}6!.$ **21.** $\frac{13!}{2!2!2!} - \binom{3}{1}\frac{12!}{2!2!} + \binom{3}{2}\frac{11!}{2!} - 10!.$
22. $\sum_{k=0}^{3}(-1)^k\binom{4}{k}(4-k)^{20}.$ **23.** The coefficient of $z^n/n!$ in
$\left(\frac{z^0}{0!} + \frac{z^1}{1} + \frac{z^2}{2!} + \frac{z^3}{3!}\right)\left(\frac{z^0}{0!} + \frac{z^1}{1} + \frac{z^2}{2!}\right)^2 (1+z)^5.$
24. The coefficient of z^{30} in $(z^1 + z^2 + z^3 + z^4 + z^5 + z^6)^{10}.$
25. The coefficient of z^{20} in
$[1 + 2z + 3z^2 + 4z^3 + 5z^4 + 6z^5]\binom{2}{1}\frac{1}{1-z^2}\frac{z}{1-z^2}.$

Chapter 4. Groups

§31. Symmetry Groups.

Exercise 1. Filling in the table by columns is much easier than by rows. For each column after the first, except the σ-column, all we have to do is multiply the entries in the immediate left column by ρ on the right, which is easy. Your paper square will help you get the σ-column. See Table 2.

D_4	ι	ρ	ρ^2	ρ^3	σ	$\sigma\rho$	$\sigma\rho^2$	$\sigma\rho^3$
ι	ι	ρ	ρ^2	ρ^3	σ	$\sigma\rho$	$\sigma\rho^2$	$\sigma\rho^3$
ρ	ρ	ρ^2	ρ^3	ι	$\sigma\rho^3$	σ	$\sigma\rho$	$\sigma\rho^2$
ρ^2	ρ^2	ρ^3	ι	ρ	$\sigma\rho^2$	$\sigma\rho^3$	σ	$\sigma\rho$
ρ^3	ρ^3	ι	ρ	ρ^2	$\sigma\rho$	$\sigma\rho^2$	$\sigma\rho^3$	σ
σ	σ	$\sigma\rho$	$\sigma\rho^2$	$\sigma\rho^3$	ι	ρ	ρ^2	ρ^3
$\sigma\rho$	$\sigma\rho$	$\sigma\rho^2$	$\sigma\rho^3$	σ	ρ^3	ι	ρ	ρ^2
$\sigma\rho^2$	$\sigma\rho^2$	$\sigma\rho^3$	σ	$\sigma\rho$	ρ^2	ρ^3	ι	ρ
$\sigma\rho^3$	$\sigma\rho^3$	σ	$\sigma\rho$	$\sigma\rho^2$	ρ	ρ^2	ρ^3	ι

TABLE 2. Cayley Table for D_4.

Exercise 2. We have

$(\gamma \circ (\beta \circ \alpha))(P) = \gamma((\beta \circ \alpha)(P))$ (by the definition of the composition of γ following $\beta \circ \alpha$)

$= \gamma(\beta(\alpha(P)))$ (by the definition of the composition of β following α)

$= (\gamma \circ \beta)(\alpha(P))$ (by the definition of the composition of γ following β)

$= ((\gamma \circ \beta) \circ \alpha)(P)$ (by the definition of the composition of $\gamma \circ \beta$ following α).

Since the mappings $\gamma(\beta\alpha)$ and $(\gamma\beta)\alpha$ act on each point in the same way, then the mappings are equal. This completes the proof. Note that the associative law was not used in this proof; of course, this must be so since that is exactly what we were trying to prove in the first place. Now, after the proof, we can write $\alpha\beta\gamma$ without fear of confusion, since either way of associating the factors gives the same result. Generalizing, we can associate in any way any product of elements in a group (The generalization is obvious, but its proof is a bother.)

Exercise 3. Since there are 8 vertices from which to hang a cube; since then 1 of 3 of the vertices adjacent to the top vertex can be pointed toward yourself; and since this determines the position of all the vertices of the cube, then we see that there are 24 rotation symmetries of the cube. Similar arguments can be made using either the 12 edges or else the 6 faces. The following list has 24 different rotation symmetries; the list must contain all the rotation symmetries of the cube. (It is always nice to know when we can stop looking.)

1 identity,

6 rotations of $\pm90°$ about the axes that join centers of opposite faces,

3 rotations of $180°$ about the axes that join centers of opposite faces,

6 rotations of $180°$ about the axes that join centers of opposite edges,

8 rotations of $\pm120°$ about the axes that join opposite vertices.

It is the rotations of $\pm120°$ that are evasive. Without holding a cube in your hand with the cube in the position described above as hanging, it is difficult to see that a cube has a three-fold symmetry. For most of us, this is hard to discern with the cube in "normal" position, sitting on a face. With the cube in the unlikely position of being balanced on 1 vertex, the three-fold symmetry is evident.

Exercise 4. The regular dodecahedron has $12 \cdot 5$ rotation symmetries. Thus, when we find 60 symmetries, we can stop looking for more.

1 identity,

6·4 rotations of $\pm 72°$ or $\pm 144°$ about the axes that join centers of opposite faces,

15·1 rotations of $180°$ about the axes that join centers of opposite edges,

10·2 rotations $\pm 120°$ about the axes that join opposite vertices.

§32. Legendre's Theorem.

Exercise 1. Suppose α, β, and γ are in group G and $\gamma \odot \alpha = \gamma \odot \beta$. Then, by the inverse property, $\gamma^{-1} \odot (\gamma \odot \alpha) = \gamma^{-1} \odot (\gamma \odot \beta)$. Next, by the associative property, $(\gamma^{-1} \odot \gamma) \odot \alpha = (\gamma^{-1} \odot \gamma) \odot \beta$. By the inverse property again, $\iota \odot \alpha = \iota \odot \beta$. Finally, by the identity property, $\alpha = \beta$, as desired. Whether we just proved the left cancellation law or the right cancellation law depends on what country you come from. Americans call this one the left cancellation law. The proof of the other cancellation law is similar.

Exercise 2. The expressions

$$(z^{-1}y^{-1}x^{-1}\cdots d^{-1}c^{-1}b^{-1}a^{-1})(abcd\cdots xyz)$$

and

$$(abcd\cdots xyz)(z^{-1}y^{-1}x^{-1}\cdots d^{-1}c^{-1}b^{-1}a^{-1})$$

can be rewritten, after innumerable uses of the associative law, as

$$z^{-1}(y^{-1}(x^{-1}(\cdots(c^{-1}(b^{-1}(a^{-1}a)b)c)\cdots)x)y)z)$$

and

$$a(b(c(\cdots(x(y(zz^{-1})y^{-1})x^{-1})\cdots)c^{-1})b^{-1})a^{-1},$$

each of which implodes to the identity. Hence, by the definition of an inverse element, we have

$$(abcd\cdots xyz)^{-1} = z^{-1}y^{-1}x^{-1}\cdots d^{-1}c^{-1}b^{-1}a^{-1}.$$

$$
\begin{array}{c|cccccc}
S_3 &
\begin{pmatrix}1&2&3\\1&2&3\end{pmatrix} &
\begin{pmatrix}1&2&3\\2&3&1\end{pmatrix} &
\begin{pmatrix}1&2&3\\3&1&2\end{pmatrix} &
\begin{pmatrix}1&2&3\\1&3&2\end{pmatrix} &
\begin{pmatrix}1&2&3\\3&2&1\end{pmatrix} &
\begin{pmatrix}1&2&3\\2&1&3\end{pmatrix} \\
\hline
\begin{pmatrix}1&2&3\\1&2&3\end{pmatrix} &
\begin{pmatrix}1&2&3\\1&2&3\end{pmatrix} &
\begin{pmatrix}1&2&3\\2&3&1\end{pmatrix} &
\begin{pmatrix}1&2&3\\3&1&2\end{pmatrix} &
\begin{pmatrix}1&2&3\\1&3&2\end{pmatrix} &
\begin{pmatrix}1&2&3\\3&2&1\end{pmatrix} &
\begin{pmatrix}1&2&3\\2&1&3\end{pmatrix} \\
\begin{pmatrix}1&2&3\\2&3&1\end{pmatrix} &
\begin{pmatrix}1&2&3\\2&3&1\end{pmatrix} &
\begin{pmatrix}1&2&3\\3&1&2\end{pmatrix} &
\begin{pmatrix}1&2&3\\1&2&3\end{pmatrix} &
\begin{pmatrix}1&2&3\\2&1&3\end{pmatrix} &
\begin{pmatrix}1&2&3\\1&3&2\end{pmatrix} &
\begin{pmatrix}1&2&3\\3&2&1\end{pmatrix} \\
\begin{pmatrix}1&2&3\\3&1&2\end{pmatrix} &
\begin{pmatrix}1&2&3\\3&1&2\end{pmatrix} &
\begin{pmatrix}1&2&3\\1&2&3\end{pmatrix} &
\begin{pmatrix}1&2&3\\2&3&1\end{pmatrix} &
\begin{pmatrix}1&2&3\\3&2&1\end{pmatrix} &
\begin{pmatrix}1&2&3\\2&1&3\end{pmatrix} &
\begin{pmatrix}1&2&3\\1&3&2\end{pmatrix} \\
\begin{pmatrix}1&2&3\\1&3&2\end{pmatrix} &
\begin{pmatrix}1&2&3\\1&3&2\end{pmatrix} &
\begin{pmatrix}1&2&3\\3&2&1\end{pmatrix} &
\begin{pmatrix}1&2&3\\2&1&3\end{pmatrix} &
\begin{pmatrix}1&2&3\\1&2&3\end{pmatrix} &
\begin{pmatrix}1&2&3\\2&3&1\end{pmatrix} &
\begin{pmatrix}1&2&3\\3&1&2\end{pmatrix} \\
\begin{pmatrix}1&2&3\\3&2&1\end{pmatrix} &
\begin{pmatrix}1&2&3\\3&2&1\end{pmatrix} &
\begin{pmatrix}1&2&3\\2&1&3\end{pmatrix} &
\begin{pmatrix}1&2&3\\1&3&2\end{pmatrix} &
\begin{pmatrix}1&2&3\\3&1&2\end{pmatrix} &
\begin{pmatrix}1&2&3\\1&2&3\end{pmatrix} &
\begin{pmatrix}1&2&3\\2&3&1\end{pmatrix} \\
\begin{pmatrix}1&2&3\\2&1&3\end{pmatrix} &
\begin{pmatrix}1&2&3\\2&1&3\end{pmatrix} &
\begin{pmatrix}1&2&3\\1&3&2\end{pmatrix} &
\begin{pmatrix}1&2&3\\3&2&1\end{pmatrix} &
\begin{pmatrix}1&2&3\\2&3&1\end{pmatrix} &
\begin{pmatrix}1&2&3\\3&1&2\end{pmatrix} &
\begin{pmatrix}1&2&3\\1&2&3\end{pmatrix} \\
\end{array}
$$

TABLE 3. Cayley Table for S_3.

S_3	(1)	(123)	(132)	(23)	(13)	(12)
(1)	(1)	(123)	(132)	(23)	(13)	(12)
(123)	(123)	(132)	(1)	(12)	(23)	(13)
(132)	(132)	(1)	(123)	(13)	(12)	(23)
(23)	(23)	(13)	(12)	(1)	(123)	(132)
(13)	(13)	(12)	(23)	(132)	(1)	(123)
(12)	(12)	(23)	(13)	(123)	(132)	(1)

TABLE 4. Cayley Table for S_3, Redux.

§33. Permutation Groups.

Exercise 1. See Table 3.

Exercise 2. See Table 4.

Exercise 3. $(abcd\ldots yz) = (az)(ay)\ldots(ad)(ac)(ab)$; and
$(ab)(ab) = (abc)(acb)$, $(ab)(ac) = (acb)$, and otherwise
$(ab)(cd) = (cad)(abc)$.

Exercise 4. Even: (1), $(12)(34)$, $(13)(24)$, $(14)(23)$, (123), (132), (124), (142), (134), (143), (234), (243). Multiply each of these on the left by any odd permutation in S_n, say (12), to get the odd elements: (12), (34), (1324), (1423), (23), (13), (24), (14), (1342), (1432), (1234), (1243). Since we have 4! elements, we must have all of them. Also, $|A_n| = n!/2$ since half the elements in S_n are even and half are odd. (It follows that if a permutation group has any odd elements then half of the permutations are odd.) Now we show that A_n is a subgroup of S_n. Since mappings are associative, A_n has the associative property. Since the sum of 2 even integers is an even integer, then A_n is closed. Since $(12)(12) = (1)$, then the identity is even and A_n has the identity property. Since a transposition is its own inverse and since the inverse of product is the product of the inverses in reverse order, then A_n has the inverse property.

§34. Generators.

Exercise. Group A_4, as does every group, has 1 subgroup of order 1: $\langle(1)\rangle$. The cosets here are just the 12 1-element subsets. The other trivial subgroup is A_4, which has 1 coset, itself. Group A_4 has 3 subgroups of order 2: $\langle(12)(34)\rangle$, $\langle(13)(24)\rangle$, and $\langle(14)(23)\rangle$. With $H = \langle(12)(34)\rangle$, then the 6 cosets of H are given by $H = \{(1), (12)(34)\}$, $(13)(24)H = \{(13)(24), (14)(23)\}$, $(123)H = \{(123), (134)\}$, $(132)H = \{(132), (234)\}$, $(124)H = \{(124), (143)\}$, and $(142)H = \{(142), (243)\}$. Group A_4 has 1 subgroup of order 4: If $H = \{(1), (12)(34), (13)(24), (14)(23)\}$, then the 3 cosets of H are given by H, $(123)H = \{(123), (134), (243), (142)\}$ and $(132)H = \{(132), (234), (124), (143)\}$. Here, for example, note that $(132)H = (234)H = (124)H = (143)H$. Group A_4 has 4 subgroups of order 3: $\langle(123)\rangle$, $\langle(124)\rangle$, $\langle(134)\rangle$, and $\langle(234)\rangle$. With $H = \langle(123)\rangle$, the 4 cosets of H are given by $H = \{(1), (123), (132)\}$, $(124)H = \{(124), (14)(23), (134)\}$,

$(142)H = \{(142), (234), (13)(24)\}$, and
$(243)H = \{(243), (143), (12)(34)\}$.
Group A_4 has no subgroup of order 6.

§35. Cyclic Groups.

Homework. We have $G = C_{12}$, with $n = 12 = 2^2 3$ and $g = \rho$, where $C_{12} = \langle \rho \rangle$.

k	$\gcd(n,k)$	$\langle g^k \rangle$ has order $\frac{n}{\gcd(n,k)}$.
1	1	$\langle \rho \rangle$ has order 12.
2	2	$\langle \rho^2 \rangle$ has order 6.
3	3	$\langle \rho^3 \rangle$ has order 4.
4	4	$\langle \rho^4 \rangle$ has order 3.
5	1	$\langle \rho^5 \rangle$ has order 12.
6	6	$\langle \rho^6 \rangle$ has order 2.
7	1	$\langle \rho^7 \rangle$ has order 12.
8	4	$\langle \rho^8 \rangle$ has order 3.
9	3	$\langle \rho^9 \rangle$ has order 4.
10	2	$\langle \rho^{10} \rangle$ has order 6.
11	1	$\langle \rho^{11} \rangle$ has order 12.
12	12	$\langle \rho^{12} \rangle$ has order 1.

Further:

Divisor d of 12	$\phi(d)$	Subgroup of order d
1	1	$\langle \rho^{12} \rangle = \{\rho^0\}$
2	1	$\langle \rho^6 \rangle = \{\rho^6, \rho^0\}$
3	2	$\langle \rho^4 \rangle = \langle \rho^8 \rangle = \{\rho^4, \rho^8, \rho^0\}$
4	2	$\langle \rho^3 \rangle = \langle \rho^9 \rangle = \{\rho^3, \rho^6, \rho^9, \rho^0\}$
6	2	$\langle \rho^2 \rangle = \langle \rho^{10} \rangle = \{\rho^2, \rho^4, \rho^6, \rho^8, \rho^{10}, \rho^0\}$
12	4	$\langle \rho \rangle = \langle \rho^5 \rangle = \langle \rho^7 \rangle = \langle \rho^{11} \rangle = C_{12}$
	sum: 12	

§36. Equivalence and Isomorphism.

Exercise 1. $\boxed{a} = \{x \in S \mid x \sim a\}$. Suppose $s \in \boxed{a} \cap \boxed{c}$. Then, $s \sim a$ and $s \sim c$. So, $a \sim s$ and $s \sim c$. Hence, $a \sim c$ and, likewise, $c \sim a$. Thus, $x \sim a$ implies $x \sim c$. Likewise, $x \sim c$ implies $x \sim a$. Therefore, $\boxed{a} = \boxed{c}$.

Homework. 1. Since $D_3 = \langle \rho, \sigma \rangle$, it follows that we know where every element of D_3 goes if we know where each of ρ and σ goes under the mapping. Four isomorphisms, π, λ, μ, and ν, are determined by Table 5. (Can you find the other 2 isomorphisms?)

α	$\pi(\alpha)$	$\lambda(\alpha)$	$\mu(\alpha)$	$\nu(\alpha)$
ι	(1)	(1)	(1)	(1)
ρ	(123)	(123)	(132)	(132)
ρ^2	(132)	(132)	(123)	(123)
σ	(12)	(23)	(12)	(13)
$\sigma\rho$	(23)	(13)	(13)	(12)
$\sigma\rho^2$	(13)	(12)	(23)	(23)

TABLE 5. Isomorphisms from D_3 to S_3.

2. Since $C_{12} = \langle \rho \rangle = \langle \rho^5 \rangle$, the mapping π determines a one-to-one correspondence. In particular, ρ^{5r} is mapped to ρ^r, since $(\rho^{5r})^5 = \rho^{25r} = \rho^{12 \cdot 2r + r} = (\rho^{12})^{2r} \rho^r = \rho^r$. Also,

$$\pi(\rho^t \rho^s) = \pi(\rho^{t+s}) = \rho^{5(t+s)} = \rho^{5t+5s} = \rho^{5t} \rho^{5s} = \pi(\rho^t)\pi(\rho^s),$$

as desired. **3.** Since,

$$\pi(\rho^t \rho^s) = \pi(\rho^{t+s}) = \rho^{4(t+s)} = \rho^{4t+4s} = \rho^{4t} \rho^{4s} = \pi(\rho^t)\pi(\rho^s),$$

then π is a homomorphism. We know that π is not an isomorphism because there is no k for which $\rho^{4k} = \rho$.

How many groups of order 2 are there?

If G is a subgroup of group I and α is in I, then $\{\alpha g \alpha^{-1} \mid g \in G\}$ is easily checked to be a subgroup of I. This subgroup is denoted by $\alpha G \alpha^{-1}$ and is called the **conjugate of G by α**. If G and H are subgroups of group I, then H **is conjugate to G in I** if there is an element α is in I such that $H = \alpha G \alpha^{-1}$. Conjugation is an equivalence relation on the subgroups of I.

For the plane, let ρ be the 180° rotation about the origin O, and let σ be the reflection in h, the X-axis. So, $C_2 = \{\iota, \rho\}$ and $D_1 = \{\iota, \sigma\}$. Now, it turns out that the conjugate of ρ by isometry α is the 180° rotation about the point $\alpha(O)$. Further, the conjugate of σ by isometry α is the reflection in the line $\alpha(h)$. Therefore, all the groups of order 2 consisting of the identity and a 180° rotation

Isometry g	π_g in cycle notation	Cycle type of π_g
ι	(A)	z_1^6
ρ	$(ACE)(BDF)$	z_3^2
ρ^2	$(AEC)(BFD)$	z_3^2
$\sigma_{\overleftrightarrow{AD}}$	$(BF)(CE)$	$z_1^2 z_2^2$
$\sigma_{\overleftrightarrow{BE}}$	$(AC)(DF)$	$z_1^2 z_2^2$
$\sigma_{\overleftrightarrow{CF}}$	$(AE)(BD)$	$z_1^2 z_2^2$

TABLE 6. D_3 Acting on Vertices of a Hexagon.

are conjugate to each other, and all the groups of order 2 consisting of the identity and a reflection in a line are conjugate to each other. These are the only subgroups or order 2. Since the conjugate of a reflection is always a reflection and never a rotation of $180°$, it follows that the groups C_2 and D_1 cannot be conjugate. The geometer wants to distinguish these 2 groups as different since they have very different geometric properties, even though they are isomorphic as abstract groups. For this reason, the geometer says that there are 2 plane isometry groups of order 2. In the isometries of 3-space, there are 3 groups of order 2 (up to conjugation).

Chapter 5. Actions

§37. The Definition.

Exercise 1. We first observe that, in the group of all permutations on X, mapping π_e is the identity mapping on X, by axiom 1 in the definition. Mapping π_g is onto since $\pi_g(\pi_{g^{-1}}(x)) = x$ and is 1–to–1 since $\pi_g(x) = \pi_g(y)$ implies
$x = \pi_e(x) = \pi_{g^{-1}}(\pi_g(x)) = \pi_{g^{-1}}(\pi_g(y)) = \pi_e(y) = y$. So, π_g is a permutation on X. Then, since $\pi_g \pi_{g^{-1}} = \pi_{gg^{-1}} = \pi_e$ and $\pi_{g^{-1}} \pi_g = \pi_{g^{-1}g} = \pi_e$, we have

$$\pi_g^{-1} = \pi_{g^{-1}}.$$

All other parts of the exercise follow from this important equation.

Homework. 1. See Table 6, where, in general, if l is a line then σ_l is the reflection in l. **2.** See Table 7. **3.** See Table 8.

Exercise 2. See Table 9 for the first part of the exercise. One

$g \in D_4$	Part (a) π_g & type	Part (b) π_g & type	Part (c) π_g & type	Part (d) π_g & type
ι	(1) & z_1^4	(a) & z_1^4	(x) & z_1^2	(a) & z_1^6
ρ	(1234) & z_4^1	$(abcd)$ & z_4^1	(xy) & z_2^1	$(abcd)(xy)$ & $z_2^1 z_4^1$
ρ^2	$(13)(24)$ & z_2^2	$(ac)(bd)$ & z_2^2	(x) & z_1^2	$(ab)(cd)$ & $z_1^2 z_2^2$
ρ^3	(1423) & z_4^1	$(adcb)$ & z_4^1	(xy) & z_2^1	$(abcd)(xy)$ & $z_2^1 z_4^1$
σ_h	(24) & $z_1^2 z_2^1$	$(ad)(bc)$ & z_2^2	(x) & z_1^2	$(ad)(bc)$ & $z_1^2 z_2^2$
σ_v	(13) & $z_1^2 z_2^1$	$(ab)(cd)$ & z_2^2	(x) & z_1^2	$(ab)(cd)$ & $z_1^2 z_2^2$
σ_p	$(12)(34)$ & z_2^2	(bd) & $z_1^2 z_2^1$	(xy) & z_2^1	$(bd)(xy)$ & $z_1^2 z_2^2$
σ_m	$(14)(23)$ & z_2^2	(ac) & $z_1^2 z_2^1$	(xy) & z_2^1	$(ac)(xy)$ & $z_1^2 z_2^2$

TABLE 7. D_4 Acting on Various Sets.

$g \in D_4$	π_g	Cycle type
ι	(1)	z_1^{16}
ρ	$(1)(2,5,4,3)(6,8,11,9)(7,10,)(12,15,14,13)(16)$	$z_1^2 z_2^1 z_4^3$
ρ^2	$(1)(2,4)(3,5)(6,11)(7)(8,9)(10)(12,14)(13,15)(16)$	$z_1^4 z_2^6$
ρ^3	$(1)(2,3,4,5)(6,9,11,8)(7,10)(12,13,14,15)$	$z_1^2 z_2^1 z_4^3$
σ_h	$(1)(2,5)(3,4)(6,11)(7,10)(8)(9)(12,15)(13,14)(16)$	$z_1^4 z_2^6$
σ_v	$(1)(2,3)(4,5)(6)(7,10)(8,9)(11)(12,13)(14,15)(16)$	$z_1^4 z_2^6$
σ_p	$(1)(2,4)(3)(5)(6,9)(7)(8,11)(10)(12,14)(13)(15)(16)$	$z_1^8 z_2^4$
σ_m	$(1)(2)(3,5)(4)(6,8)(7)(9,11)(10)(12)(13,15)(14)(16)$	$z_1^8 z_2^4$

TABLE 8. D_4 Acting on Colorings of the 2–by–2 Checkerboard.

G	e	a	b	c
e	e	a	b	c
a	a	e	c	b
b	b	c	e	a
c	c	b	a	e

Π	(e)	$(ea)(bc)$	$(eb)(ac)$	$(ec)(ab)$
(e)	(e)	$(ea)(bc)$	$(eb)(ac)$	$(ec)(ab)$
$(ea)(bc)$	$(ea)(bc)$	(e)	$(ec)(ab)$	$(eb)(ac)$
$(eb)(ac)$	$(eb)(ac)$	$(ec)(ab)$	(e)	$(ea)(bc)$
$(ec)(ab)$	$(ec)(ab)$	$(eb)(ac)$	$(ea)(bc)$	(e)

TABLE 9. Cayley Tables for G and Π.

$$
\begin{array}{c|cccc}
G & e & a & b & c \\
\hline
e & e & a & b & c \\
a & a & e & & \\
b & b & & e & \\
c & c & & & e
\end{array}
$$

TABLE 10. Cayley Table for Noncyclic Group of Order 4.

X	$\{1, 2, 3, 4\}$	$\{a, b, c, d\}$	$\{x, y\}$	$\{a, b, c, d\ x, y\}$	
x	$1, 2, 3, 4$	a, b, c, d	x, y	a, b, c, d	x, y
$\lvert O_x \rvert$	4	4	2	4	2
$\lvert S_x \rvert$	2	2	4	2	4

TABLE 11.

possibility for a group of order 4 is a cyclic group, generated by an element of order 4. All such groups are isomorphic to C_4. On the other hand, if there are no elements of order 4, then every element except the identity must be of order 2. If we assume that the elements are e, a, b, c with e the identity, then the Cayley table must have the structure exhibited in Table 10. However, neither of the products ab nor ba can be equal to any of e, a, or b, and so in each case the product must be c. Then, the rest of the table is also uniquely determined.

§38. Burnside's Lemma.

Exercise. See Tables 11 and 12.

Homework.
1. $\omega = (2^{64} + 2^{16} + 2^{32} + 2^{16})/4$.
2. $\omega = (2^{64} + 2^{16} + 2^{32} + 2^{16} + 2^{32} + 2^{32} + 2^{36} + 2^{36})/8$.
3. $\omega = (3^{64} + 3^{16} + 3^{32} + 3^{16} + 3^{32} + 3^{32} + 3^{36} + 3^{36})/8$.
4. $\omega = (m^{64} + m^{16} + m^{32} + m^{16} + m^{32} + m^{32} + m^{36} + m^{36})/8$.

X		2–Colorings of Fixed 2-by-2 Checkerboard				
x	1	2, 3, 4, 5	6, 8, 9, 11	7, 10	12, 13, 14, 15	16
$\lvert O_x \rvert$	1	4	4	2	4	1
$\lvert S_x \rvert$	8	2	2	4	2	8

TABLE 12.

5. $\omega = (m^{49} + m^{13} + m^{25} + m^{13} + m^{28} + m^{28} + m^{28} + m^{28})/8$.

6. Note that it does not hurt to write the coefficient 1 in a cycle index. This may make it easier to check that all the coefficients in the numerator add up to $|G|$, the denominator. Here

$$\omega = \frac{\overbrace{1 \cdot 3^6}^{0°} + \overbrace{2 \cdot 3^1}^{\pm 60°} + \overbrace{2 \cdot 3^2}^{\pm 120°} + \overbrace{1 \cdot 3^3}^{180°} + \overbrace{3 \cdot 3^4}^{diag.} + \overbrace{3 \cdot 3^3}^{other}}{12} = 92.$$

7. The first 5 of the terms in the numerator below come from consideration of the rotations of $0°$, $\pm 45°$, $\pm 90°$, $\pm 135°$, and $180°$, respectively. The last 2 terms are from reflections in a diagonal and the reflections in the perpendicular bisector of an "edge." We have

$$\omega = \frac{1 \cdot 4^8 + 2 \cdot 4^1 + 2 \cdot 4^2 + 2 \cdot 4^1 + 1 \cdot 4^4 + 4 \cdot 4^5 + 4 \cdot 4^4}{16} = 4435.$$

§39. Applications of Burnside's Lemma.

Homework 1.

1. Cycle index for the rotations of a cube acting on the *faces*:
The identity rotation ι with

$\pi_\iota = (1)(2)(3)(4)(5)(6)$ $\Rightarrow 1[z_1^6]$.

6 rotations of order 4, like α with

$\pi_\alpha = (1265)(3)(4)$ $\Rightarrow 6[z_1^2 z_4^1]$.

3 rotations of order 2, like α^2 with

$\pi_{\alpha^2} = (16)(25)(3)(4)$ $\Rightarrow 3[z_1^2 z_2^2]$.

6 rotations of order 2, like β with

$\pi_\beta = (14)(25)(36)$ $\Rightarrow 6[z_2^3]$.

8 rotations of order 3, like γ with

$\pi_\gamma = (124)(365)$ $\Rightarrow 8[z_3^2]$.

We have considered all 24 rotations of the cube and have cycle index

$$\frac{z_1^6 + 3z_1^2 z_2^2 + 6z_1^2 z_4 + 6z_2^3 + 8z_3^2}{24}.$$

2. Cycle index for the rotations of a cube acting on the *edges*: Here

$\pi_\iota = (a)(b)(c)(d)(e)(f)(g)(h)(i)(j)(k)(l)$ $\Rightarrow 1[z_1^{12}]$.

$\pi_\alpha = (abcd)(efgh)(ijkl)$ $\Rightarrow 6[z_4^3]$.

$\pi_{\alpha^2} = (ac)(bd)(eg)(fh)(ik)(jl)$ $\Rightarrow 3[z_2^6]$.

$\pi_\beta = (af)(b)(ce)(dj)(gi)(hk)(l)$ $\Rightarrow 6[z_1^2 z_2^5].$
$\pi_\gamma = (ajg)(bfc)(dek)(hil)$ $\Rightarrow 8[z_3^4].$
We have considered all 24 rotations of the cube and have cycle index

$$\frac{z_1^{12} + 3z_2^6 + 6z_1^2 z_2^5 + 8z_3^4 + 6z_4^3}{24}.$$

3. Cycle index for the rotations of a cube acting on the *vertices*:
Here
$\pi_\iota = (A)(B)(C)(D)(E)(F)(G)(H)$ $\Rightarrow 1[z_1^8].$
$\pi_\alpha = (ABCD)(EFGH)$ $\Rightarrow 6[z_4^2].$
$\pi_{\alpha^2} = (AC)(BD)(EG)(FH)$ $\Rightarrow 3[z_2^4].$
$\pi_\beta = (AG)(BC)(DF)(EH)$ $\Rightarrow 6[z_2^4].$
$\pi_\gamma = (AFH)(BGD)(C)(E)$ $\Rightarrow 8[z_1^2 z_3^2].$
We have considered all 24 rotations of the cube and have cycle index

$$\frac{z_1^8 + 9z_2^4 + 8z_1^2 z_3^2 + 6z_4^2}{24}.$$

Homework 2.

1. "It's *déjà vu* all over again." Let m be 5 in #3 above.

2. Since this particular choice of letters allows a reflection through the middle letter of the 17-letter word to produce a word in these letters, we can think of the group as D_1 containing the identity isometry and a reflection. However, the solution is independent of the letters given. If the given letters were A, B, C, D, for example, then the action is still a group of order 2. Here, X is the set of 17-letter words containing the given letters and G consisting of the identity mapping on all words and the mapping that reverses the spelling of each word. Earlier in our study, to solve this problem, we would simply have added the number of palindromes (words whose reverse spelling is the same as the word) to the number of ordered words and divided the sum by 2. We would unknowingly have been using an elementary application of Burnside's Lemma to get

$$\frac{\frac{17!}{2!4!5!6!} + \frac{8!}{1!2!2!3!}}{2}.$$

3. We could do this problem the first week, if not before. We suppose that red, orange, and yellow are among the colors. We place the face of the cube that is to be red flat on the table. There are 5 ways to color the top face. If the top face is orange, then color the front face yellow; if the top face is not orange, then color the front face orange. In each of the 5 cases case, we have 3 ordered faces remaining and 3 colors not yet used. There are 3! ways to color these faces with the colors. Thus, there are $5 \cdot 3!$ ways to color the cube as required.

From another view, there are 6! ways to color the faces of a fixed cube with the 6 colors, if each color is used, and there are 24 ways of rotating the cube, since 24 is the order of the octahedral group. Thus, there are 6!/24 ways to color the cube as required. This solution is tantamount to using Burnside's Lemma, which gives us

$$\frac{[1]6! + [23]0}{24}.$$

4. Considering the 6 elements in D_3, from the array

$$
\begin{array}{lll}
\iota & [1] \; z_1^{10} & [1] \; z_1^{15} \\
\rho, \rho^2 & [2] \; z_1^1 z_3^3 & [2] \; z_3^5 \\
\sigma_1, \sigma_2, \sigma_3 & [3] \; z_1^2 z_2^4 & [3] \; z_1^3 z_2^6
\end{array}
$$

we get our solutions

$$\frac{m^{10} + 3m^6 + 2m^4}{6} \quad \text{and} \quad \frac{m^{15} + 3m^9 + 2m^5}{6}.$$

5. We have

$$\omega^\star = \frac{\overbrace{1 \cdot 3^{60}}^{0^\circ} + \overbrace{2 \cdot 3^{15}}^{\pm 90^\circ} + \overbrace{1 \cdot 3^{30}}^{180^\circ} + \overbrace{2 \cdot 3^{33}}^{h,v} + \overbrace{2 \cdot 3^{30}}^{p,m}}{8}.$$

Not that anybody asked, but $\omega = 9$ here.

6. If the figure consisting of the 18 edges is fixed by a rotation, then the cube itself is fixed by the rotation. (We should not expect the converse, however.) So we run through the elements of the octahedral group to see which of these rotations actually fix the marked cube. Of course, there is the identity rotation ι and π_ι has type z_1^{18}. Since the rotations of $\pm 90^\circ$ about the join of centers of opposite faces do not fix the marked cube, they are irrelevant. However, each of the 3 rotations of 180° about the joins of the centers of opposite faces is associated with a permutation of type

$z_1^2 z_2^6$. Rotations of $180°$ about the joins of the centers of opposite edges do not fix the marked cube. The $4 \cdot 2$ rotations of $\pm 120°$ about the join of opposite vertices have an associated permutation of type z_3^6. We have

$$\omega^\star = \frac{[1]5^{18} + [3]5^8 + [8]5^6}{12}$$

as our answer.

The 12 elements of the group of rotations of the marked cube are those of the **tetrahedral group**, which is the group of all rotations, including the identity, that fix a given regular tetrahedron. To check this out, while holding a cube by 2 opposite vertices, on each of 3 faces mark a diagonal emanating from the top vertex. Then join the other ends of these diagonals in pairs. We now have each face of the cube marked with a diagonal. These marked diagonals are the edges of a regular tetrahedron, and this regular tetrahedron is inscribed in the cube.

§40. Pólya's Pattern Inventory.

Homework.

1. The pattern inventory of the black and white colorings of the vertices of a cube is given in the text immediately before this problem as

$$b^8 w^0 + b^7 w^1 + 3b^6 w^2 + 3b^5 w^3 + 7b^4 w^4 + 3b^3 w^5 + 3b^2 w^6 + b^1 w^7 + b^0 w^8.$$

We see that $b^4 w^4$ has coefficient 7, which is the answer to our question.

2. The cycle index for D_4 acting on the 4 squares of the 2–by–2 checkerboard is

$$\frac{[1]z_1^4 + [2]z_4^1 + [1]z_2^2 + [2]z_2^2 + [2]z_1^2 z_2^1}{8}.$$

The pattern inventory for the 2-coloring is then

$$\frac{(b+w)^4 + 2(b^4 + w^4) + 3(b^2 + w^2)^2 + 2(b+w)^2(b^2 + w^2)}{8},$$

which is

$$b^4 + b^3 w + 2b^2 w^2 + b w^3 + w^4,$$

which should be no surprise.

3. From the cycle index above (§39, Homework 1, #1, page 209), we compute the pattern inventory

$$[(b + w)^6 + 3(b + w)^2(b^2 + w^2)^2$$
$$+ 6(b + w)^2(b^4 + w^4) + 6(b^2 + w^2)^3$$
$$+ 8(b^3 + w^3)^2] /24,$$

which is

$$1b^6 + 1b^5w + 2b^4w^2 + 2b^3w^3 + 2b^2w^4 + 1bw^5 + 1w^6.$$

If we think about it, we could have written down the simplified pattern inventory without the first form above, since each of the coefficients in the simplified form is easy to obtain by elementary counting.

4. From the cycle index above (§39, Homework 1, #2, page 209) we get the pattern inventory

$$[(r + w + b)^{12} + 3(r^2 + w^2 + b^2)^6$$
$$+ 6(r + w + b)^2(r^2 + w^2 + b^2)^5 + 8(r^3 + w^3 + b^3)^4$$
$$+ 6(r^4 + u^4 + b^4)^3] /24.$$

5. We want the coefficient of $r^4w^4b^4$ in the expansion of the pattern inventory in #4 above. The coefficient of $r^4w^4b^4$ in $(r^2 + w^2 + b^2)^6$ is the coefficient of $(r^2)^2(w^2)^2(b^2)^2$ in $((r^2) + (w^2) + (b^2))^6$. The calculation of the coefficient of $r^4w^4w^4$ in $(r + w + b)^2(r^2 + w^2 + b^2)^5$ is more formidable. Think of selecting 1 term from each of the 7 factors to get a term in the expansion. We must pick 1 of the 3 choices r, w, b from the first factor. Say, r. Then we are forced to pick r again from the second factor, since we can pick up only even exponents from the remaining 5 factors. So, we now have r^2 accounted for and need to select another r^2 from 1 of the 5 last factors. Next, we must pick w^2 from each of 2 of the remaining 4 factors, and, finally, we must pick b^2 from each of the remaining 2 factors. Thus, as a final answer, we get

$$\frac{\frac{12!}{4!4!4!} + 3\frac{6!}{2!2!2!} + 6\left[3 \cdot \binom{5}{1}\binom{4}{2}\binom{2}{2}\right] + 8 \cdot 0 + 6\frac{3!}{1!1!1!}}{24},$$

or 1479.

For a slightly different approach to the calculation of the quantity in the square brackets above, we expand $(r + w + b)^2$ as $r^2 + w^2 + b^2 + J$, where, here, J is the junk $2rw + 2rb + 2br$. (There is danger in discarding the wrong thing in the junk. If we were looking for the coefficient of $r^3w^5b^4$, then this junk would be a jewel.) However, it is evident that our junk can make no contribution to the coefficient of $r^4w^4w^4$. In fact, the junk can be ignored here, and we are essentially looking for the coefficient of $r^4w^4w^4$ in $(r^2 + w^2 + b^2)^6$, which is 90.

Homework for a Week.

1. By now, it should almost be as easy to compute the cycle index

$$\frac{z_1^8 + 6z_4^2 + 3z_2^4 + 6z_2^4 + 8z_1^2z_3^2}{24}$$

as it is to look it up in the text. Thus, the pattern inventory is

$$\big[(r + w + b)^8$$
$$+ 6(r^4 + w^4 + b^4)^2 + 9(r^2 + w^2 + b^2)^4$$
$$+ 8\,(r + w + b)^2(r^3 + w^3 + b^3)^2\big] /24.$$

2. The cycle index is, in this case,

$$\frac{z_1^{64} + 2z_4^{16} + 3z_2^{32} + 2z_1^8z_2^{28}}{8}$$

and so the pattern inventory here is

$$\big[(b + w)^{64} + 2(b^4 + w^4)^{16} + 3(b^2 + w^2)^{32}$$
$$+ 2\,(b + w)^8(b^2 + w^2)^{28}\big] /8.$$

We first find s, the number of colorings where there are an equal number of black and white squares. The first 3 terms in the numerator above should not be a problem. For example, the coefficient of $b^{32}w^{32}$ in $(b^4 + w^4)^{16}$ is the coefficient of $(b^4)^8(w^4)^8$ in $((b^4) + (w^4))^{16}$. This coefficient is $16!/(8!8!)$. However, the last term in the numerator might give us a problem. Now, from $(b + w)^8$ we can get any one of b^8w^0, b^6w^2, b^4w^4, b^2w^6, or b^0w^8, if we ignore the irrelevant terms having an odd power of b. Thus, respectively, we must get $b^{24}w^{32}$, $b^{26}w^{30}$, $b^{28}w^{28}$, $b^{30}w^{26}$, and $b^{32}w^{24}$ from $(b^2 + w^2)^{28}$. Therefore, we have a calculation for s.

$$s = \frac{1}{8}\binom{64}{32} + \frac{2}{8}\binom{16}{8} + \frac{3}{8}\binom{32}{16} +$$
$$\frac{2}{8}\left[\binom{8}{8}\binom{28}{12} + \binom{8}{6}\binom{28}{13} + \binom{8}{4}\binom{28}{14} + \binom{8}{2}\binom{28}{13} + \binom{8}{0}\binom{28}{12}\right].$$

Next, we compute t, the total number of 2-colorings. But that is easy. We get

$$t = \frac{2^{64} + 2^{17} + 3 \cdot 2^{32} + 2^{37}}{8}.$$

The solution to our problem is

$$\frac{t - s}{2}.$$

3. The cycle index appears in #1 above. We want the coefficient of $r^2 w^2 b^2 g^2$ in the related pattern inventory. The solution is

$$\frac{1}{24}\left[\frac{8!}{2!2!2!2!} + 6 \cdot 0 + 9\frac{4!}{1!1!1!1!} + 8 \cdot 0\right],$$

or 114.

4. In each of the cases below, we let w_3 denote the number of necklaces with exactly 3 white beads.

Case $n = 7$. The cycle index is

$$\frac{z_1^7 + 6z_7^1 + 7z_1^1 z_2^3}{14}.$$

Case $n = 7$ and $k = 2$. The pattern inventory is

$$\left[(b + w)^7 + 6(b^7 + w^7) + 7(b + w)(b^2 + w^2)^3\right]/14$$

and so

$$w_3 = \left[\binom{7}{3} + 0 + 7(1 \cdot 3)\right]/14 = 4.$$

Case $n = 7$ and $k = 3$. The pattern inventory is

$$\left[(r + w + b)^7 + 6(r^7 + w^7 + b^7) + 7(r + w + b)(r^2 + w^2 + b^2)^3\right]/14$$

and so

$$w_3 = \left[\binom{7}{3} 2^4 + 0 + 7 \left(1 \cdot \binom{3}{1} 2^2 \right) \right] / 14 = 46.$$

Case $n = 9$. The cycle index is

$$\frac{z_1^9 + 2z_3^3 + 6z_9^1 + 9z_1^1 z_2^4}{18}.$$

Case $n = 9$ and $k = 2$. The pattern inventory is

$$\left[(b+w)^9 + 2(b^3 + w^3)^3 + 6(b^9 + w^9) \right.$$
$$\left. + 9(b+w)(b^2 + w^2)^4 \right] / 18$$

and so

$$w_3 = \left[\binom{9}{3} + 2 \cdot 3 + 0 + 9 \cdot 4 \right] / 18 = 7.$$

Case $n = 9$ and $k = 3$. The pattern inventory is

$$\left[(r+w+b)^9 + 2(r^3 + w^3 + b^3)^3 \right.$$
$$\left. + 6(r^9 + w^9 + b^9) + 9\,(r+w+b)(r^2 + w^2 + b^2)^4 \right] / 18$$

and so

$$w_3 = \left[\binom{9}{3} 2^6 + 2(3 \cdot 2^2) + 0 + 9 \left(\binom{4}{1} 2^3 \right) \right] / 18 = 316.$$

Case $n = 11$. The cycle index is

$$\frac{z_1^{10} + 10z_{11}^1 + 11z_1^1 z_2^5}{22}.$$

Case $n = 11$ and $k = 2$. The pattern inventory is

$$\left[(b+w)^{11} + 10(b^{11} + w^{11}) + 11(b+w)(b^2 + w^2)^5 \right] / 22$$

and so

$$w_3 = \left[\binom{11}{3} + 0 + 11 \cdot 1 \cdot \binom{5}{1} \right] / 22 = 10.$$

Case $n = 11$ and $k = 3$. The pattern inventory is

$$[(r + w + b)^{11}$$
$$+ 10(r^{11} + w^{11} + b^{11})$$
$$+ 11(r + w + b)(r^2 + w^2 + b^2)^5] / 22$$

and so

$$w_3 = \left[\binom{11}{3} 2^8 + 0 + 11 \cdot 1 \cdot \binom{5}{1} 2^4\right] / 22 = 1960.$$

If we had a choice, would we rather do a necklace problem that had 36 beads or one that had 37 beads? No contest. Actually, for any odd prime p we should be able to write down the cycle index for D_p acting on the regular p-gon without much hesitation.

5. The cycle index here is

$$\frac{z_1^{16} + 2z_4^4 + z_2^8}{4}.$$

By using the Lemma to PIE, we have our solution

$$\frac{3^{16} + 2 \cdot 3^4 + 3^8}{4} - \binom{3}{2}\frac{2^{16} + 2 \cdot 2^4 + 2^8}{4} + \binom{3}{1}\frac{1 + 2 + 1}{4}.$$

6. From the computations in the last 2 columns of

Type of g in S_3	# of type	Part a	Part b
z_1^3	1	$\left\langle \genfrac{}{}{0pt}{}{3}{12} \right\rangle$	$\left\langle \genfrac{}{}{0pt}{}{3}{14} \right\rangle \left\langle \genfrac{}{}{0pt}{}{3}{4} \right\rangle^2$
$z_1 z_2$	3	7	$8 \cdot 3 \cdot 3$
z_3	2	1	0

we have the solutions

$$\frac{\left\langle \genfrac{}{}{0pt}{}{3}{12} \right\rangle + 3 \cdot 7 + 2 \cdot 1}{6} = 19$$

and

$$\frac{\left\langle \genfrac{}{}{0pt}{}{3}{14} \right\rangle \left\langle \genfrac{}{}{0pt}{}{3}{4} \right\rangle^2 + 3(8 \cdot 3 \cdot 3) + 2 \cdot 0}{6} = 4536$$

for the first and second questions, respectively.

order d	$\# = \phi(d)$	g, for example	cycle type of π_g	$\|F_g^\star\|$
1	1	ι	z_1^{28}	$2^{28} + 2$
2	1	ρ^{14}	z_2^{14}	$2^{14} + 2$
4	2	ρ^7	z_4^7	$2^7 - 2$
7	6	ρ^4	z_7^4	$2^4 + 2$
14	6	ρ^2	z_{14}^2	$3 \cdot 2$
28	12	ρ^1	z_{28}^1	0
14 reflections in diagonals			$z_1^2 z_2^{13}$	$3 \cdot 2^{14}$
14 other reflections			z_2^{14}	0

TABLE 13. D_{28} Acting on Necklace.

§41. Necklaces.

Homework.

1. We use the lemma. The first 4 of the terms in the numerator below come from consideration of the rotations of $0°$, $90°$, $180°$, and $270°$, respectively. The last 2 terms are from reflections in the 2 diagonals and the reflections in the 2 perpendicular bisectors of the edges. We have

$$\omega^\star = \frac{(2^4 + 2) + 0 + (2^2 + 2) + 0 + 2(3 \cdot 2^2) + 2(0)}{8} = 6.$$

Without the lemma, we can consider the number of necklaces with exactly 0 red, 1 red, or 2 red beads. These are easy to count directly; we get 1, 2, 3, respectively.

2. From Table 13, where ρ is a rotation of order 28, we get the computation

$$\omega^\star = \frac{(2^{28} + 2) + (2^{14} + 2) + 2(2^7 - 2) + 6(2^4 + 2) + 6(3 \cdot 2) + 12(0) + 14(3 \cdot 2^{14}) + 14(0)}{56}$$

$$= 4,806,078.$$

3. First note that the center vertex can be any 1 of the 5 colors. Since this vertex is adjacent to all others, we are then reduced to 4-coloring an octagon with 4 colors such that no 2 adjacent vertices have the same color. From a table similar to Table 13, we get the computation

$$\omega^\star =$$

$$5 \times \frac{\overbrace{1(3^8+3)}^{0^\circ}+\overbrace{1(3^4+3)}^{180^\circ}+\overbrace{2(4\cdot 3)}^{\pm 90^\circ}+\overbrace{4(0)}^{\pm 45^\circ,\pm 135^\circ}+\overbrace{4(4\cdot 3^4)}^{diag.}+\overbrace{4(0)}^{other}}{16}$$

$$= 2490.$$

Practice Exam. 1. 75. **2.** 75. It is no accident that the answers to #1 and #2 are the same. If we join the centers of adjacent faces of a tetrahedron, we get another tetrahedron. **3.** 278. **4.** 704,370.

Tidbit. Stirling number of the first kind $\begin{bmatrix} r \\ \ell n \end{bmatrix}$, which counts the number of ways to seat n persons at k indistinguishable round tables with at least 1 person at each table, is the number of permutations in S_n with exactly k cycles.

Chapter 6. Recurrence Relations

§42. Examples of Recurrence Relations.

Exercise. From the bottom up: $a_0 = 0$, $a_1 = 2a_0 + 2 = 2$,
$a_2 = 2a_1 + 2 = 2^2 + 2$, $a_3 = 2a_2 + 2 = 2(2^2 + 2) + 2 = 2^3 + 2^2 + 2$,
$a_4 = 2a_3 + 2 = 2^4 + 2^3 + 2^2 + 2, \ldots$,
$a_n = 2^n + 2^{n-1} + \cdots + 2^3 + 2^2 + 2 = 2^{n+1} - 2$. From the top down:
$a_n = 2a_{n-1} + 2 = 2[2a_{n-2} + 2] + 2 = 2^2 a_{n-2} + 2^2 + 2 =$
$2^2[2a_{n-3} + 2] + 2^2 + 2 = 2^3 2a_{n-3} + 2^3 + 2^2 + 2 = \cdots =$
$2^{n-1} 2a_{n-(n-1)} + 2^{n-1} + \cdots + 2^2 + 2 = 2^n + 2^{n-1} + \cdots + 2^2 + 2 = 2^{n+1} - 2$.

§43. The Fibonacci Numbers.

MORE PIE PROBLEMS?
Line Pizza. No problem here.

$$c_n = n + 1 \text{ for } n \geq 0.$$

Plane Pizza. The maximality property requires that every 2 lines intersect in a point but no 3 lines intersect at 1 point. Since $n - 1$ lines cut the plane into b_{n-1} regions and an nth line will intersect

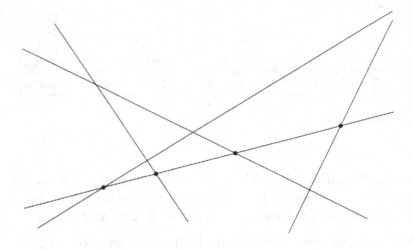

FIGURE 2. Pizza Slices.

the $n-1$ lines in $n-1$ different points, then this n^{th} line will create a_{n-1} new regions, as is seen in Figure 2. That is, $b_n = b_{n-1} + a_{n-1}$ for $n > 0$ with $b_0 = 1$. So $b_n = b_{n-1} + n$ and, unwinding, we get $b_n = b_0 + [1 + 2 + +3 + \cdots + (n-1) + n]$. Therefore,

$$b_n = \frac{n^2 + n + 2}{2} \text{ for } n \geq 0.$$

Space Pizza. The maximality property requires that every 2 planes intersect in a line but no 3 planes intersect at 1 line. An nth plane will intersect $n-1$ planes in $n-1$ lines such that each 2 of these lines intersect in a point in the nth plane but no 3 of these lines intersect at 1 point in the nth plane. We can now view Figure 2 as the cross section determined by the nth plane cutting the space pizza slices determined by $n-1$ planes. Since $n-1$ planes cut space into a_{n-1} regions and the nth plane creates b_{n-1} new regions, we have $a_n = a_{n-1} + b_{n-1}$ for $n > 0$ with $a_0 = 1$. Using the fact that $\sum_{i=0}^{r} i^2 = r(r+1)(2r+1)/6$ and unwinding, we get $a_n = a_0 + \sum_{k=0}^{n-1} b_k = 1 + \sum_{k=0}^{n-1} \frac{k^2 + k + 2}{2}$
$= 1 + \frac{1}{2}[\sum_{k=0}^{n-1} k^2 + \sum_{k=0}^{n-1} k + \sum_{k=0}^{n-1} 2]$
$= 1 + \frac{1}{2}\left[\frac{(n-1)(n)(2n-1)}{6} + \frac{(n-1)(n)}{2} + 2n \right]$
$= \frac{(n+1)(n^2-n+6)}{6}$. So

$$c_n = \frac{n^3 + 5n + 6}{6} \text{ for } n \geq 0.$$

It is interesting that
$$c_n = \binom{n}{0} + \binom{n}{1},$$
$$b_n = \binom{n}{0} + \binom{n}{1} + \binom{n}{2},$$
$$a_n = \binom{n}{0} + \binom{n}{1} + \binom{n}{2} + \binom{n}{3}.$$

§44. A Dozen Recurrence Problems.

The Dozen. 1. The nth circle intersects the other $n - 1$ circles in $2(n - 1)$ points that determine $2(n - 1)$ arcs on the nth circle and each arc increases the number of regions by 1. So, $a_n = a_{n-1} + 2(n - 1)$ for $n > 1$ with $a_1 = 2$. If we want to declare that $a_0 = 1$, then we must do so separately because the recurrence relation fails for $n = 1$. **2.** If there are k spaces left after parking the first car, then we can fill the remaining spaces with the cars in a_k ways. (That's what "a_k" means!) Thus, considering the color of the first that is parked and how much room is left for the remaining cars, we get $a_n = a_{n-1} + 2a_{n-2}$ for $n > 2$ with $a_1 = 1$ and $a_2 = 3$. (Perhaps you would like to include $a_0 = 1$?) **3.** If there are k stairs left to climb, then we can climb these stairs in a_k ways. Considering the number of stairs taken in the first step and how many stairs are left for the remaining steps, we get $a_n = a_{n-1} + a_{n-2}$ for $n > 2$ with $a_1 = 1$ and $a_2 = 2$. **4.** The sequence can begin $01 \ldots$ in a_{n-2} ways, begin $02 \ldots$ in a_{n-2} ways, begin $1 \ldots$ in a_{n-1} ways, and begin $2 \ldots$ in a_{n-1} ways. So, $a_n = 2a_{n-1} + 2a_{n-2}$ for $n > 1$ with $a_0 = 1$ and $a_1 = 3$. (Check $a_2 = 3^2 - 1$.) **5.** Considering whether n is in the subset or not, we get $a_n = a_{n-1} + a_{n-2}$ for $n > 2$ with $a_1 = 2$ and $a_2 = 3$. Can you explain why this is the same problem as Example 14? **6.** Considering the color of the top flag, we get $a_n = 2a_{n-1} + a_{n-2}$ for $n > 1$ with $a_0 = 1$ and $a_1 = 2$. **7.** We have $a_n - a_{n-1} = 2[a_{n-1} - a_{n-2}]$, or $a_n = 3a_{n-1} - 2a_{n-2}$, for $n > 1$ with $a_0 = 2$ and $a_1 = 7$. **8.** Remember that n is the dollar amount. For each purchase, we spend either \$1 in 1 of 2 ways or else \$2 in 1 of 3 ways. Hence, $a_n = 2a_{n-1} + 3a_{n-2}$ for $n > 1$ with $a_0 = 1$ and $a_1 = 2$. (Check $a_2 = 7$.) **9.** From the figure

R........	a_{n-1}	GR........	a_{n-2}
W........	a_{n-1}	GW........	a_{n-2}
B........	a_{n-1}	GB........	a_{n-2}

we see that $a_n = 3a_{n-1} + 3a_{n-2}$ for $n > 0$ with $a_0 = 1$ and $a_1 = 4$. (Check $a_2 = 4^2 - 1$.) **10.** The second head can appear on the nth

flip in $n - 1$ ways. Otherwise, the nth flip can be either heads or tails together with the second head appearing on or before the $(n - 1)^{\text{st}}$ flip in $2a_{n-1}$ ways. So $a_n = 2a_{n-1} + n - 1$ for $n > 0$ with $a_0 = 0$. Unwinding here would be difficult. A better approach is to consider all possibilities for the n flips. We get 0 heads, 1 head, or at least 2 heads. So, $1 + n + a_n = 2^n$, and we therefore have $a_n = 2^n - n - 1$ for $n \geq 0$. **11.** We find the maximum M_1 and minimum m_1 among half of the 2^n reals in a_{n-1} ways. Likewise, we find the maximum M_2 and minimum m_2 among the other 2^{n-1} reals in a_{n-1} ways. It is necessary and sufficient to compare M_1 with M_2 and to compare m_1 with m_2. So, $a_n = 2a_{n-1} + 2$ for $n > 1$ with $a_1 = 1$. If we want to declare that $a_0 = 0$, we have to do so separately because the argument here fails for $n = 0$. Unwinding produces $a_n = 3 \cdot 2^{n-1} - 2$ for $n > 0$. **12.** We have $a_{r,n} = 0$ if $r < 2n$ or $r > 4n$, $a_{2,1} = 6$, $a_{3,1} = 10$, $a_{4,1} = 15$, and $a_{4,2} = 36$. Otherwise we have $a_{r,n} = \binom{3}{2}a_{r-2,n-1} + \binom{3}{3}a_{r-3,n-1} + \binom{3}{4}a_{r-4,n-1}$, from considering how many balls we put in the first box.

§45. Solving Recurrence Relations.

Homework. 1. Unwinding, we get $a_n = a_1 + [2(1) + 2(2) + \cdots + 2(n - 1)]$. So, $a_n = n^2 - n + 2$ for $n > 0$. **2.** We first substitute x^k for a_k in the recurrence relation $a_n = a_{n-1} + 2a_{n-2}$. From $x^n = x^{n-1} + 2x^{n-2}$ we then get the characteristic equation $x^2 - x - 2 = (x - 2)(x + 1) = 0$, which has roots -1 and 2. So $a_n = b(-1)^n + c(2)^n$ is the general solution, where b and c are arbitrary constants. Using the initial conditions, we have the equations $1 = a_1 = -b + 2c$ and $3 = a_2 = b + 4c$, which have solution $b = 1/3$ and $c = 2/3$. So, $a_n = [2^{n+1} + (-1)^n]/3$ for $n \geq 1$. **3.** $a_n = F_{n+1}$ for $n \geq 0$. **4.** From $x^n = 2x^{n-1} + 2x^{n-2}$ we get the characteristic equation $x^2 - 2x - 2 = 0$, which has roots $1 \pm \sqrt{3}$. So $a_n = b(1 + \sqrt{3})^n + c(1 - \sqrt{3})^n$ is the general solution. Using the initial conditions, we have the equations $1 = a_0 = b + c$ and $3 = a_1 = b(1 + \sqrt{3}) + c(1 - \sqrt{3})$. (In passing, we note that the empty sequence really pays off here. The first of the 2 equations above is fairly simple and much preferable to using the next possibility, which is $8 = a_2 = b(1 + \sqrt{3})^2 + c(1 - \sqrt{3})^2$.) This time we have the solutions $b = (3 + 2\sqrt{3})/6$ and $c = (3 - 2\sqrt{3})/6$. So $a_n = \frac{(3+2\sqrt{3})(1+\sqrt{3})^n + (3-2\sqrt{3})(1-\sqrt{3})^n}{6}$ for $n \geq 0$. **5.** $a_n = F_{n+2}$ for $n \geq 1$. **6.** From $x^n = 2x^{n-1} + x^{n-2}$ we get the characteristic

equation $x^2 - 2x - 1 = 0$, which has roots $1 \pm \sqrt{2}$. So
$a_n = b(1 + \sqrt{2})^n + c(1 - \sqrt{2})^n$ is the general solution. Using the
initial conditions, we have the equations $1 = a_0 = b + c$ and
$2 = a_1 = b(1 + \sqrt{2}) + c(1 - \sqrt{2})$. We have the solutions
$b = (2 + \sqrt{2})/4$ and $c = (2 - \sqrt{2})/4$. So
$a_n = \frac{(2+\sqrt{2})(1+\sqrt{2})^n+(2-2\sqrt{2})(1-\sqrt{2})^n}{4}$ for $n \geq 0$. **7.** From
$x^n = 3x^{n-1} - 2x^{n-2}$ we get the characteristic equation
$x^2 - 3x + 2 = 0$, which has roots 1 and 2. So $a_n = b(1)^n + c(2)^n$ is
the general solution. Using the initial conditions, we have the
equations $2 = a_0 = b + c$ and $7 = a_1 = b + 2c$, which have solution
$b = -3$ and $c = 5$. So, $a_n = 5 \cdot 2^n - 3$ for $n \geq 0$. **8.** From
$x^n = 2x^{n-1} + 3x^{n-2}$ we get the characteristic equation
$x^2 - 2x - 3 = 0$, which has roots -1 and 3. So $a_n = b(-1)^n + c(3)^n$
is the general solution. Using the initial conditions, we have the
equations $1 = a_0 = b + c$ and $2 = a_1 = -b + 3c$, which have solution
$b = 1/4$ and $c = 3/4$. So, $a_n = \frac{3^{n+1}+(-1)^n}{4}$ for $n \geq 0$.

§46. The Catalan Numbers.

Modeling Problems. 1. If we drop the first element of such a
sequence we have such a sequence. Thinking of what could come
before such a sequence of length $n - 1$, we see that $a_n = 2a_{n-1}$ for
$n > 2$ with $a_1 = 3$ and $a_2 = 6$. So $a_n = 2^{n-1}3$ for $n > 0$. Note that
$a_0 = 1$. **2.** By considering, in turn, sequences that begin $1 \ldots, 2 \ldots,$
$0 \ldots$, we see that $a_n = a_{n-1} + a_{n-1} + (3^{n-1} - a_{n-1})$. So
$a_n = a_{n-1} + 3^{n-1}$ for $n > 0$ with $a_0 = 1$. By unwinding, we get
$a_n = \frac{1+3^n}{2}$ for $n \geq 0$. **3.** $a_n = (2n - 1)a_{n-1}$ for $n > 0$ with $a_0 = 1$.
The $2n - 1$ is the number of ways to match the tallest (say) with
someone. Unwinding here gives $(2n - 1)(2n - 3) \cdots (5)(3)(1)$, or
$(2n)!/[2^n n!]$, as expected. **4.** $a_n = \binom{4n-1}{3}a_{n-1}$ for $n > 0$ with
$a_0 = 1$. Unwinding here gives the expected solution $(4n)!/[4!^n n!]$.
5. Considering the piles that from bottom up begin RG $\ldots$,
W $\ldots$, B $\ldots$, and G $\ldots$, with the obvious notation, we see that
$a_n = 3a_{n-1} + a_{n-2}$ for $n > 1$ with $a_0 = 1$ and $a_1 = 3$. (Check
$a_2 = 10$.) Piles of poker chips of n possible colors are n-ary
sequences in party dress. **6.** Either the second head occurs before
the nth flip in a_{n-1} ways or else on the nth flip in $n - 1$ possible
ways. So, $a_n = a_{n-1} + (n - 1)$ for $n > 1$ with $a_1 = 0$. Therefore,
$a_n = n(n - 1)/2$ for $n \geq 0$. **7.** Taking into consideration whether
the sequence begins with 1, 2, 3, or 0, we see that

$a_n = a_{n-1} + a_{n-1} + a_{n-1} + (4^{n-1} - a_{n-1})$. So, $a_n = 2a_{n-1} + 4^{n-1}$ for $n > 1$ with $a_1 = 3$ or, if you prefer, $a_n = 2a_{n-1} + 4^{n-1}$ for $n > 0$ with $a_0 = 1$. Unwinding produces $a_n = 2^{n-1}(2^n - 1)$ for $n \geq 0$.
8. We can attach a 0 to the front of every such sequence of length $n - 1$ except those that begin with a 1. So the number of desired sequences that begin with 0 is $a_{n-1} - a_{n-2}$. We can attach a 1 or a 2 to the front of every such sequence of length $n - 1$ to obtain a desired sequence. All desired sequences are obtained in these ways. We have $a_n = 3a_{n-1} - a_{n-2}$ for $n > 1$ with $a_0 = 1$ and $a_1 = 3$. (Check $a_2 = 8$.) **9.** The sequence can begin $01\ldots$ in a_{n-2} ways, begin $02\ldots$ in a_{n-2} ways, begin $1\ldots$ in a_{n-1} ways, and begin $2\ldots$ in a_{n-1} ways. So, $a_n = 2a_{n-1} + 2a_{n-2}$ for $n > 1$ with $a_0 = 1$ and $a_1 = 3$. (Check $a_2 = 3^2 - 1$.) **10.** Taking into consideration whether the sequence begins with 1, 2, or 0, we see that $a_n = a_{n-1} + a_{n-1} + 2^{n-1}$. So, $a_n = 2a_{n-1} + 2^{n-1}$ for $n > 0$ with $a_0 = 1$. (Check $a_1 = 3$.) **11.** The sequence can begin $00\ldots$ in 3^{n-2} ways, begin $01\ldots$ in a_{n-2} ways, begin $02\ldots$ in 3^{n-2} ways, begin $1\ldots$ in a_{n-1} ways, and begin $2\ldots$ in a_{n-1} ways. So, $a_n = 2a_{n-1} + a_{n-2} + 2 \cdot 3^{n-2}$ for $n > 1$ with $a_0 = 0$ and $a_1 = 1$. (Check $a_2 = 4$.) **12.** The sequence can begin $0\ldots$ in 0 ways, begin $1\ldots$ in 4^{n-1} ways, begin $2\ldots$ in a_{n-1} ways, and begin $3\ldots$ in a_{n-1} ways. So, $a_n = 2a_{n-1} + 4^{n-1}$ for $n > 0$ with $a_0 = 0$. The following attack gives the solution in closed form. Each of the k elements after the first 1 can be anything and the $(n - 1) - k$ elements before this 1 can be only 2's and 3's. So, $a_n = \sum_{k=0}^{n-1} 4^k 2^{n-1-k} = 2^{n-1} \sum_{k=0}^{n-1} 2^k = 2^{n-1}(2^n - 1)$ for $n \geq 0$. An analogous argument gives a_n in Example 15. **13.** The figures ⌐, ⊟, and ⊏ show the 3 ways that the tiles can be placed to cover an end of the board having length 2. We therefore see that $a_n = a_{n-1} + 2a_{n-2}$ for $n > 1$ with $a_0 = a_1 = 1$. (Check $a_2 = 3$.)
14a. From the bottom, a pile can begin with a white chip in a_{n-1} ways, begin with a blue chip in a_{n-1} ways, begin with a green chip in a_{n-1} ways, but begin with a red chip in $a_{n-1} - a_{n-2}$ ways. We arrive at the last count by considering all such piles of height $n - 1$ sitting on top of a red chip; all of these pile of height $n - 1$ can be used except those that begin with a green chip. Conversely, every such pile of height n that begins with a red chip must be of this form. So, $a_n = 4a_{n-1} - a_{n-2}$ for $n > 1$ with $a_0 = 1$ and $a_1 = 4$.

(Check $a_2 = 2^4 - 1$.) **14b.** We resort to the figure

RR........	4^{n-2}	W........	a_{n-1}
RW........	4^{n-2}	B........	a_{n-1}
RB........	4^{n-2}	G........	a_{n-1}
RG........	a_{n-2}		

So, $a_n = 3a_{n-1} + a_{n-2} + 3 \cdot 4^{n-2}$ for $n > 1$ with $a_0 = 0$ and $a_1 = 1$.
(Check $a_2 = 6$.) **14c.** The top chip can be red in a_{n-1} ways. The
top chip can be white in a_{n-1} ways. The top chip can be blue in
a_{n-1} ways. The top chip can be green in $4^{n-1} - 3^{n-1}$ ways. So,
$a_n = 3a_{n-1} + 4^{n-1} - 3^{n-1}$ for $n > 0$ with $a_0 = 0$. (Check $a_1 = 0$.)
On the other hand, the number of possible piles such that the
bottom red chip is the kth from the bottom is $3^{k-1} \cdot (4^{n-k} - 3^{n-k})$.
So, $a_n = \sum_{k=1}^{n} 3^{k-1}(4^{n-k} - 3^{n-k}) = 4^{n-1} \sum_{k=1}^{n} (3/4)^{k-1} -$
$3^{n-1} \sum_{k=1}^{n} 1 = 4^n - 3^n - n3^{n-1} = 4^n - (n-3)3^{n-1}$ for $n \geq 0$.
14d. The top chip can be red in 0 ways. The top chip can be white
in a_{n-1} ways. The top chip can be blue in a_{n-1} ways. The top chip
can be green in 4^{n-1} ways. So, $a_n = 2a_{n-1} + 4^{n-1}$ for $n > 0$ with
$a_0 = 1$. (Check $a_1 = 3$.) **14e.** The top chip can be red in 4^{n-1} ways.
The top chip can be white in a_{n-1} ways. The top chip can be blue
in a_{n-1} ways. The top chip can be green in 0 ways. So,
$a_n = 2a_{n-1} + 4^{n-1}$ for $n > 0$ with $a_0 = 0$. (Check $a_1 = 1$.) **15.** Given
a "desirable" sequence of n H's and n T's, we index the H's and
index the T's, as in the example where the desirable sequence
$HTHHTHTT$ gives the indexed sequence $H_1T_1H_2H_3T_2H_4T_3T_4$.
Note that the indexed sequence not only has H_{i+1} following H_i and
T_{i+1} following T_i but also T_i following H_i. Now to our $2n$ persons
arranged in order of increasing heights, we first assign the terms of
the indexed sequence in order. Then we form the array such that
the persons assigned the H_i's are in the first row in increasing order
of the index i. Likewise, the persons assigned the T_i's are in the
second row in increasing order of the index i. That T_i follows H_i in
the sequence assures us that columns are also arranged in
increasing order of height. Conversely, for a given array, assign each
of the persons in the first row an H and each person in the second
row a T. Then form a sequence of H's and T's by arranging all $2n$
persons in a line in increasing heights. Argue that the sequence of n
H's and n T's is a desirable sequence. Hence, our answer is C_n.

Practice Exam. 1. The sequence can begin 0 in a_{n-1} ways, begin 1 in 3^{n-1} ways, begin 2 in a_{n-1} ways, and begin 3 in a_{n-1} ways. So, $a_n = 3a_{n-1} + 3^{n-1}$ for $n > 0$ with $a_0 = 1$. (Check $a_1 = 4$.)

2. The sequence can begin 0 in a_{n-1} ways, begin 10 in a_{n-2} ways, begin 2 in a_{n-1} ways, and begin 3 in a_{n-1} ways. So, $a_n = 3a_{n-1} + a_{n-2}$ for $n > 1$ with $a_0 = 1$ and $a_1 = 3$. (Check, $a_2 = 10$.)

3. From the figure

$00\ldots$	4^{n-2}		$1\ldots$	a_{n-1}
$01\ldots$	a_{n-2}		$2\ldots$	a_{n-1}
$02\ldots$	a_{n-2}		$3\ldots$	a_{n-1}
$03\ldots$	a_{n-2}			

we get, $a_n = 3a_{n-1} + 3a_{n-2} + 4^{n-2}$ for $n > 1$ with $a_0 = a_1 = 0$. (Check $a_2 = 1$.)

4. If the first 2 digits of a desired sequence of length n are different, then the sequence is obtained by attaching to the front of a sequence of length $n - 1$ any of 3 possible digits. On the other hand, if the first 2 digits are the same, then they must be 2's or 3's and followed by a desired sequence of length $n - 2$. We also note that cutting off digits from an end of any desired sequence produces a desired sequence. So, $a_n = 3a_{n-1} + 2a_{n-2}$ for $n > 1$ with $a_0 = 1$ and $a_1 = 4$. (Check $a_2 = 4^2 - 2$.) This very clever solution is also obtained by first producing the recurrence relations $z_n = z_{n-1} + 2a_{n-2}$ and $a_n = 2z_{n-1} + 2a_{n-1}$, where z_n counts the number of desired sequences of length n that begin with a 0 and thus z_n also counts the number of desired sequences of length n that begin with a 1.

5. Let z_n be the number of such sequences of length n that begin with 0. The number beginning with 1 is also z_n. The number beginning with 2 is a_{n-1}. The number beginning with 3 is a_{n-1}. So, $a_n = 2z_n + 2a_{n-1}$, or $2z_n = a_n - 2a_{n-1}$. From the figure

$00\ldots\ldots\ldots$	z_{n-1}
$01\ldots\ldots\ldots$	0
$02\ldots\ldots\ldots$	a_{n-2}
$03\ldots\ldots\ldots$	a_{n-2}

we see that $z_n = z_{n-1} + 2a_{n-2}$, or $2z_n = 2z_{n-1} + 4a_{n-2}$. Hence, substituting, we now have $[a_n - 2a_{n-1}] = [a_{n-1} - 2a_{n-2}] + 4a_{n-2}$. So, $a_n = 3a_{n-1} + 2a_{n-2}$ for $n > 1$ with $a_0 = 1$ and $a_1 = 4$. (Check $a_2 = 4^2 - 2$.) It is somewhat interesting that this problem has exactly the same answer as the previous problem. Given a sequence that satisfies either of the problems and interchanging 0 and 1 in the odd numbered terms of this sequence, we obtain a sequence that satisfies the other problem.

6. Let z_n be the number of such sequences of length n that begin with 0. So, $a_n = z_n + 3a_{n-1}$, or $z_n = a_n - 3a_{n-1}$. From the figure

$$000.......... \quad z_{n-1}$$
$$001.......... \quad a_{n-3}$$
$$002.......... \quad a_{n-3}$$
$$003.......... \quad a_{n-3}$$

we see that $z_n = z_{n-1} + 3a_{n-3}$. Hence, we now have $[a_n - 3a_{n-1}] = [a_{n-1} - 3a_{n-2}] + 3a_{n-3}$. So, $a_n = 4a_{n-1} - 3a_{n-2} + 3a_{n-3}$ for $n > 2$ with $a_0 = 1$, $a_1 = 3$, and $a_2 = 10$. (Check $a_3 = 34$.)

7. Let z_n be the number of such sequences of length n that begin with 0. Considering the second digit of these sequences that begin with a 0, we have a first relation $z_n = z_{n-1} + 3a_{n-2}$. A desired sequence can begin with a 1 (and so must begin 10) in z_{n-1} ways. A desired sequence can begin with a 2 in a_{n-2} ways. A desired sequence can begin with a 3 in a_{n-2} ways. So, we have $a_n = z_n + z_{n-1} + 2a_{n-1}$ as a second relation. Subtracting our first relation from our second, we get $2z_n = a_n - 2a_{n-1} + 3a_{n-2}$ as a third relation. Finally, using this to substitute back into the second relation, we get

$$2a_n = [a_n - 2a_{n-1} + 3a_{n-2}] + [a_{n-1} - 2a_{n-2} + 3a_{n-3}] + 2a_{n-1}.$$

So, $a_n = a_{n-1} + a_{n-2} + 3a_{n-3}$ for $n > 2$ with $a_0 = 1$, $a_1 = 3$, and $a_2 = 11$. (Check $a_3 = 39$.)

8. We will do this problem 2 different ways and get the answer in 2 different forms. We can attach a 0 to the front of every such sequence of length $n - 1$ unless that sequence begins with 00. Hence, a desired sequence can begin with a 0 in $a_{n-1} - 3a_{n-4}$ ways. A desired sequence can begin with each of a 1, 2, or 3 in a_{n-1} ways.

So we have $a_n = 4a_{n-1} - 3a_{n-4}$ for $n > 4$ with $a_0 = 1$, $a_1 = 4$, and $a_2 = 16$, $a_3 = 63$. (Check $a_4 = 249$.) A different result comes from the figure

001...	a_{n-3}	01...	a_{n-2}	1...	a_{n-1}
002...	a_{n-3}	02...	a_{n-2}	2...	a_{n-1}
003...	a_{n-3}	03...	a_{n-2}	3...	a_{n-1}

So, $a_n = 3a_{n-1} + 3a_{n-2} + 3a_{n-3}$ for $n > 2$ with $a_0 = 1$, $a_1 = 4$, and $a_2 = 16$. (Check $a_3 = 63$.)

9. A desired sequence that begins with a 0 can be formed by adding a 0 to the front of every desired sequence of length $n - 1$ except those that begin with a 01. Thus, there are $a_{n-1} - a_{n-3}$ sequences that begin with a 0. We have $a_n = 4a_{n-1} - a_{n-3}$ for $n > 2$ with $a_0 = 1$, $a_1 = 4$, and $a_2 = 16$. (Check $a_3 = 63$.)

10. From the recurrence relations $a_n = z_n + 3a_{n-1}$ and $z_n = a_{n-1} - z_{n-2}$, we end up with $a_n = 4a_{n-1} - a_{n-2} + 3a_{n-3}$ for $n > 2$ with $a_0 = 1$, $a_1 = 4$, and $a_2 = 16$. (Check $a_3 = 63$.)

§47. Nonhomogeneous Recurrence Relations.

Homework. 1a. The homogeneous part has general solution $a_n = k_1(-2)^n + k_2(3)^n$. Trying $a_n = k_3 n + k_4$ implies $(6k_3 + 3)n + (6k_4 - 13k_3) = 0$ for all n. So, $k_3 = -1/2$ and $k_4 = -13/12$, giving the general solution $a_n = k_1(-2)^n + k_2(3)^n - n/2 - 13/12$. From the initial conditions, we then get $a_n = \frac{1}{3}(-2)^n + \frac{3}{4}(3)^n - \frac{1}{2}n - \frac{13}{12}$. So, our final solution is $a_n = \frac{3^{n+2} + (-2)^{n+2} - 6n - 13}{12}$ for $n \geq 0$.

1b. The homogeneous part has general solution $a_n = k_1(-2)^n + k_2(3)^n$. Trying $a_n = k_3 2^n$ implies $k_3 = -1$, giving the general solution $a_n = k_1(-2)^n + k_2(3)^n - 2^n$. From the initial conditions, we get $a_n = \frac{4 \cdot 3^n - 2^n(5 - (-1)^n)}{5}$ for $n \geq 0$.

1c. The homogeneous part has general solution $a_n = k_1(2)^n + k_2(-1)^n$. Trying $a_n = k_3 n + k_4$ implies $k_3 = -1$ and $k_4 = -5/2$, giving the general solution $a_n = k_1(2)^n + k_2(-1)^n - n - 5/2$. From the initial conditions, we then get $k_1 = 2$ and $k_2 = 1/2$. So, our final solution is $a_n = 2^{n+1} - (n + 2) + \frac{1 - (-1)^n}{2}$ for $n \geq 0$.

1d. Trying $a_n = k_3 n 2^n$ implies $k_3 = 2/3$, giving the general solution $a_n = k_1(2)^n + k_2(-1)^n + \frac{2}{3}n 2^n$. From the initial conditions,

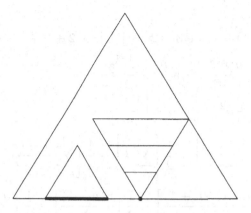

FIGURE 3.

we then get $a_n = -\frac{4}{9}(2)^n + \frac{4}{9}(-1)^n + \frac{2}{3}n2^n$. So, our final solution is
$a_n = \frac{2^{n+1}(3n-2)+4(-1)^n}{9}$ for $n \geq 0$.

1e. The homogeneous part has general solution
$a_n = k_1(2)^n + k_2n(2)^n$. Trying $a_n = k_3n^22^n$ implies $k_3 = 1/2$, giving
the general solution $a_n = k_1(2)^n + k_2n(2)^n + n^22^{n-1}$. From the
initial conditions, we then get $a_n = 2^{n-1}(n^2 - 2n + 2)$ for $n \geq 0$.

2. The number a_n will be the sum of a_{n-1} plus the number of
triangles that contain a triangle from the bottom row. These "new"
triangles with orientation $\triangle$ are easy to count because the base
determines the triangle. See Figure 3. These number
$[n + (n - 1) + \cdots + 3 + 2 + 1]$, having bases of length from 1 to n
respectively. The new triangles with orientation $\triangledown$ have a vertex on
the bottom side. These number $\underbrace{[1 + 2 + 3 + \cdots + 3 + 2 + 1]}_{n-1 \text{ terms}}$. If n is
odd, then $n - 1$ is even and this is $2\sum_{k=1}^{\frac{n-1}{2}} k$, which is $(n^2 - 1)/4$.
However, if n is even, then $n - 1$ is odd, there is a middle term, and
we have $\frac{n}{2} + \sum_{k=1}^{\frac{n-2}{2}} k$, which simplifies to $n^2/4$. In either case, this
number is given by the expression $\left[n^2 - \frac{1-(-1)^n}{2}\right]/4$. Hence, using
only simple unwinding and our last observation giving a formula for

$\sum_{k=1}^{n} k^2$, we have

$$a_n = a_{n-1} +$$
$$[n + (n-1) + \cdots + 3 + 2 + 1] + [1 + 2 + 3 + \cdots + 3 + 2 + 1]$$
$$= a_{n-1} + \frac{n(n+1)}{2} + \frac{n^2 - \frac{1-(-1)^n}{2}}{4}$$
$$= a_{n-1} + \frac{3}{4}n^2 + \frac{1}{2}n - \frac{1}{4}\left[\frac{1-(-1)^n}{2}\right]$$
$$= a_0 + \frac{3}{4}\left[\sum_{k=1}^{n} k^2\right] + \frac{1}{2}\left[\sum_{k=1}^{n} k\right] - \frac{1}{4}\left[\sum_{k=1}^{n} \frac{1-(-1)^k}{2}\right]$$
$$= \frac{3}{4}\left[\frac{n(n+1)(2n+1)}{6}\right] + \frac{1}{2}\left[\frac{n(n+1)}{2}\right] - \frac{1}{4}\left[n + \frac{1-(-1)^n}{2}\right]$$
$$= \frac{4n^3 + 10n^2 + 4n - 1 + (-1)^n}{16}$$
$$= \frac{n(n+2)(2n+1)}{8} - \frac{1-(-1)^n}{16}.$$

Chapter 7. Mathematical Induction

§48. The Principle of Mathematical Induction.

Homework MI I.
1. The formula is true when $n = 1$, as $\sum_{j=1}^{1} j^2 = 1^2 = 1 = \frac{1(2)(3)}{6}$.
For the induction step, assume that $\sum_{j=1}^{k} j^2 = \frac{k(k+1)(2k+1)}{6}$ for
some k such that $k \geq 1$. (We want to prove the formula holds for
the case $n = k + 1$.) Then, $\sum_{j=1}^{k+1} j^2 = \left[\sum_{j=1}^{k} j^2\right] + (k+1)^2 =$
$\left[\frac{k(k+1)(2k+1)}{6}\right] + \frac{6(k+1)^2}{6} = \frac{k+1}{6}[k(2k+1) + 6(k+1)] =$
$\frac{k+1}{6}(2k^2 + 7k + 6) = \frac{k+1}{6}(k+2)(2k+3) = \frac{[k+1][(k+1)+1][2(k+1)+1]}{6}$. So
the formula holds when $n = k + 1$. Thus, by mathematical
induction, the formula holds for all n such that $n \geq 1$.
2. Since we know that $\sum_{j=1}^{n} j = \frac{n(n+1)}{2}$, the formula that we are
trying to prove is $\sum_{j=1}^{n} j^3 = \left[\frac{n(n+1)}{2}\right]^2$ for $n \geq 1$. The formula is
true for $n = 1$, as $\sum_{j=1}^{1} j^3 = 1 = 1^2$. For the induction step, assume
that $\sum_{j=1}^{k} j^3 = \left[\frac{k(k+1)}{2}\right]^2$ for some k such that $k \geq 1$. (We want to

prove the formula holds for the case $n = k + 1$.) Then,

$\sum_{j=1}^{k+1} j^3 = \left[\sum_{j=1}^{k} j^3\right] + (k+1)^3 = \left(\frac{k(k+1)}{2}\right)^2 + \frac{4(k+1)^3}{4} =$

$\frac{(k+1)^2}{4}[k^2 + 4(k+1)] = \frac{(k+1)^2}{4}(k+2)^2 = \left(\frac{[k+1][(k+1)+1]}{2}\right)^2$. So the

formula holds when $n = k + 1$. Thus, by mathematical induction,
the formula holds for all n such that $n \geq 1$.

3. The formula is true for $n = 1$, as $1^2 = 1(1)(3)/3$. For the
induction step, assume that $\sum_{j=1}^{k}(2j - 1)^2 = \frac{k(2k-1)(2k+1)}{3}$ for some
k such that $k \geq 1$. (We want to prove the formula holds for the case
$n = k + 1$.) Then,

$\sum_{j=1}^{k+1}(2j - 1)^2 = \left[\sum_{j=1}^{k}(2j - 1)^2\right] + [2(k + 1) - 1]^2 =$

$\frac{k(2k-1)(2k+1)}{3} + \frac{3(2k+1)^2}{3} = \frac{2k+1}{3}[k(2k - 1) + 3(2k + 1)] =$

$\frac{2k+1}{3}(2k^2 + 5k + 3) = \frac{2k+1}{3}(k + 1)(2k + 3) = \frac{[k+1][2(k+1)-1][2(k+1)+1]}{3}$.

So the formula holds when $n = k + 1$. Thus, by mathematical
induction, the formula holds for all n such that $n \geq 1$.

4. The formula is true for $n = 1$, since $1^3 = 1 = 1(2 - 1)$. For the
induction step, assume that $\sum_{j=1}^{k}(2j - 1)^3 = k^2(2k^2 - 1)$ for some
k such that $k \geq 1$. (We want to prove the formula holds for the case
$n = k + 1$.) Then,

$\sum_{j=1}^{k+1}(2j - 1)^3 = \left[\sum_{j=1}^{k}(2j - 1)^3\right] + [2(k + 1) - 1]^3 =$

$[k^2(2k^2 - 1)] + (2k + 1)^3 = [2k^4 - k^2] + 8k^3 + 12k^2 + 6k + 1 =$

$2k^4 + 8k^3 + 11k^2 + 6k + 1 = (k^2 + 2k + 1)(2k^2 + 4k + 1) =$

$(k + 1)^2(2k^2 + 4k + 1) = (k + 1)^2[2(k + 1)^2 - 1]$. So the formula
holds when $n = k + 1$. Thus, by mathematical induction, the
formula holds for all n such that $n \geq 1$. (You may well wonder how
we were clever enough to factor the quartic polynomial in k into 2
binomials above. We weren't. Actually, we knew what the last line
should be and worked backwards to get to a quartic. Since it was
the desired quartic we got to, we knew we had an equality.)

5. First, we truly resort to induction—as opposed to mathematical
induction. We were not given a formula to prove and must devise
one of our own. We begin by computing a few of the sums, say for
$n = 1$ to $n = 5$. There is no other reasonable way to attack the
problem. We display these computations in Table 14. On this little
evidence, we now jump to the conclusion—this is what induction
is—that the formula, which we will be able to prove by the
deductive reasoning method called mathematical induction, is
$\sum_{j=1}^{n} j!j = (n + 1)! - 1$. We already know from calculating the

n	1	2	3	4	5	6
$n!$	1	2	6	24	120	720
$n!n$	1	4	18	96	600	
$\sum_{j=1}^{n} j!j$	1	5	23	119	719	

TABLE 14. Using Induction to Form a Hypothesis.

table that the basis step $(n = 1)$ follows. For the induction step, we assume that $\sum_{j=1}^{k} j!j = (k+1)! - 1$ for some k such that $k \geq 1$. Then, $\sum_{j=1}^{k+1} j!j = \left[\sum_{j=1}^{k} j!j\right] + [(k+1)!(k+1)] = [(k+1)! - 1] + [(k+1)!(k+1)] = [(k+1)!][1 + (k+1)] - 1 = [(k+1)+1]! - 1$. So the formula holds when $n = k + 1$. Thus, by mathematical induction, the formula holds for all n such that $n \geq 1$.

6. Although there are, undoubtedly, errors in this book, this one is intentional. We should ask ourself if we understand the problem. It is certainly true that if 1 die is rolled the result is either 1 of $1, 3, 5$ or else 1 of $2, 4, 6$. However, if there are 2 dice, then we are in trouble. Is the outcome 3&5 the same as 2&6? This cannot be so, since with 2 dice the 6 outcomes $2, 4, 6, 8, 10, 12$ outnumber the 5 odd outcomes $3, 5, 7, 9, 11$, if all we care about is the sum. Now, we must ask, as would be expected in a pair of dice, Are the dice indistinguishable? Again the answer must be that this cannot be the intention. We know that with 2 indistinguishable dice the number of possible outcomes is $\left\langle {6 \atop 2} \right\rangle$, which is 21. Whatever these 21 outcomes are, we are not about to find half of them even and half of them odd are we? (We can look back to Table 1 on page 184 to see the 21 outcomes.) Therefore, if the question is to make any sense, we must assume that the dice are distinguishable and proceed from there. Being asked to solve ambiguously stated problems is the kind of thing that, if not expected, should not be surprising.

Now, on to solving the revised question by mathematical induction. We conjecture that, for each positive integer n, when n distinguishable dice are rolled, the number of possible outcomes having an even sum is the same as the number of possible outcomes having an odd sum. Here, we use the notation that the number of even outcomes from rolling r distinguishable dice is E_r and that the number of odd outcomes from rolling r distinguishable dice is O_r, for each positive integer r. In particular, we have already noted that $E_1 = O_1 = 3$, which proves the basis step. For the induction step in our proof, we assume that $E_k = O_k$ for some k with $k \geq 1$ and argue that $E_{k+1} = O_{k+1}$. In order to get an even sum with $k + 1$ distinguishable dice, we need to get an even sum with the first k dice and a 2, 4, or 6 on the last die or else we need to get an odd sum on the first k dice and a 1, 3, or 5, on the last die. Thus, $E_{k+1} = 3E_k + 3O_k = 3(E_k + O_k)$. Likewise, $O_{k+1} = 3O_k + 3E_k = 3(E_k + O_k)$. Evidently, $E_{k+1} = O_{k+1}$, as desired. Thus, our conjecture follows by mathematical induction.

We may not be among the princesses of Serindip, but we have stumbled upon a rather interesting result. In the proof of the induction step above, it does not really matter whether $E_k = O_k$ is true or not. All that matters is that the last die has as many distinguishable odd sides as distinguishable even sides. This alone is sufficient to imply that $E_{k+1} = O_{k+1}$. For example, the result follows if the last die has its 6 faces numbered 1,1,1,1,1,2, regardless of what happened with all the other dice. Since the dice are distinguishable, any such die can be "the last die." Thus, we end up with the the problem that should have been asked in the first place:

Our Serendipity Problem. Prove that when n distinguishable dice, each of which can have any selection of 6 (not necessarily distinct) integers on its sides as long as there is at least 1 die with as many distinguishable even sides as distinguishable odd sides, are rolled the number of possible outcomes having an even sum equals the number of outcomes having an odd sum.

§49. The Strong Form of Mathematical Induction.

Homework MI II.

1. We want to prove that n concurrent planes such that no 3 share a line divide space into $n(n - 1) + 2$ regions for each positive integer n. The basis step is easy as 1 plane bisects space. For the induction hypothesis, we assume that k planes divide space into $k(k - 1) + 2$ regions for some k with $k \geq 1$. Now a $(k + 1)^{\text{st}}$ concurrent plane will necessarily intersect each of the k planes in a unique line. These k lines will be concurrent at the point common to all the planes. Thus, this $(k + 1)^{\text{st}}$ plane will create as many new regions as there are planar regions determined by the k lines in this plane, namely $2k$. See Figure 4, where $k = 5$. So, using the induction hypothesis, we have a total of $[k(k - 1) + 2] + 2k$ regions in space.

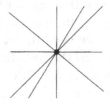

FIGURE 4. Five Concurrent Lines in a Plane.

Since $[k(k-1)+2]+2k = k^2+k+2 = (k+1)[(k+1)-1]+2$, the number of regions in space determined by $k+1$ concurrent planes is $(k+1)[(k+1)-1]+2$, as desired. Thus, assuming the formula holds for the case $n = k$ implies the formula holds for the case $n = k+1$, proving the induction step. Therefore, the formula holds for all positive integers by mathematical induction.

2. We use mathematical induction with base 8 in the basis step, which requires only observing that $8 = 3+5$. For the induction step, we assume that for some arbitrary integer k such that $k \geq 8$ there exist nonnegative integers p and q such that $k = 3p+5q$. We want to show that there are nonnegative integers r and s such that $k+1 = 3r+5s$. We now have the equalities
$k+1 = 3(p+2)+5(q-1) = 3(p-3)+5(q+2)$.
So $r = p+2$ and $s = q-1$ will do unless $q = 0$. However, in that case, we must have $p \geq 3$ since $k = 3p \geq 8$ and, therefore we have $r = p-3$ and $s = q+2$ as a solution. Thus, when $k \geq 8$, a solution for $n = k$ implies a solution for $n = k+1$. Our result now follows by mathematical induction.

3. Since $\cos 0 = 1$ and $\sin 0 = 0$, the formula holds for $n = 0$. This proves the basis step in a proof by mathematical induction with base 0. (The basis step would be even easier if the base were 1.) For the induction step, we assume that
$(\cos \alpha + i \sin \alpha)^k = \cos k\alpha + i \sin k\alpha$ for some k such that $k \geq 0$. Then, $(\cos \alpha + i \sin \alpha)^{k+1} = [\cos \alpha + i \sin \alpha]^k [\cos \alpha + i \sin \alpha] = [\cos k\alpha + i \sin k\alpha][\cos \alpha + i \sin \alpha] = [\cos k\alpha \cos \alpha - \sin k\alpha \sin \alpha] + i[\sin k\alpha \cos \alpha + \cos k\alpha \sin \alpha] = \cos(k+1)\alpha + i \sin(k+1)\alpha$. So, if the formula holds for $n = k$, then the formula holds for $n = k+1$. Hence, DeMoivre's Theorem follows by mathematical induction.

4. We have $5^2 < 2^5 < 5!$, since $25 < 32 < 120$. This proves the basis step in a proof by mathematical induction with base 5. For the induction step, we assume that $k^2 < 2^k < k!$ for some k such that $k \geq 5$. Then, $(k+1)^2 = k^2 + 2k + 1 < k^2 + k^2 < 2^k + 2^k = 2^{k+1} = 2 \cdot 2^k < 2 \cdot k! < (k+1)k! = (k+1)!$. So, $(k+1)^2 < 2^{k+1} < (k+1)!$. That the inequalities hold for the case $n = k$ implies that the inequalities hold for the case $n = k + 1$. Thus, by mathematical induction, the inequalities hold for all n such that $n \geq 5$.

5. We are given a positive integer b with $b > 1$; in the representations, b is called the *base*. For the case $0 < m < b$, we have $n = 0$ and $r_0 = m$. That is, $m = m \cdot b^0$ and there is little else to say. For the case $m \geq b$, we use the strong form of mathematical induction with base b to prove that every positive integer m can be uniquely represented in the form

$$r_0 b^0 + r_1 b^1 + r_2 b^2 + \cdots + r_{n-1} b^{n-1} + r_n b^n$$

where $r_n \neq 0$ and $0 \leq r_j < b$ for $j = 1, 2, \ldots, n$. (There's a pun there, since "base" is used in 2 different meanings—one relating to the basis step in mathematical induction and the other relating to the base in the representation of integers. By the way, we don't do the induction on n because that symbol is otherwise used.) If $m = b$, then $m = 0 + 1 \cdot b$ and there is little to prove for the basis step. For the induction step, we assume that, for some positive integer k with $k \geq b$, every positive integer less than k has a unique representation in the desired form. We want to prove that k has a unique representation in the desired form. By the so-called division algorithm (see the comments below), there are unique integers q and r such that $k = qb + r$ with $0 \leq r < b$ and $0 < q < k$. (We simply divide k by b to get quotient q and remainder r.) Since $0 < q < k$, then by the induction hypothesis there exist unique s_i such that

$q = s_0 b^0 + s_1 b^1 + s_2 b^2 + \cdots + s_{t-1} b^{t-1} + s_t b^t$ where $s_n \neq 0$ and $0 \leq s_j < b$ for $j = 1, 2, \ldots, t$. Hence, $k = r + qb$
$= r + (s_0 b^0 + s_1 b^1 + s_2 b^2 + \cdots + s_{t-1} b^{t-1} + s_t b^t)b$
$= r b^0 + s_0 b^1 + s_1 b^2 + s_2 b^3 + \cdots + s_{t-1} b^t + s_t b^{t+1}$.

We take $r_0 = r$, and we take $r_i = s_{i-1}$ for $i > 0$. Since r_0 is unique, it follows from the uniqueness of the s_i that all the r_i are also unique. So, k has a unique representation of the desired form. Therefore, our proposition follows by mathematical induction.

We have 2 comments about algorithms. First, we observe that the division algorithm is not an algorithm. What we have is a theorem that tells us about a quotient and a remainder but does not tell us how to find these. If we want to avoid misleading names, there is a greater need to change the name of mathematical induction than to change the name of the division algorithm. Our second comment concerns an application of the proof above. As a corollary of the proof we get an algorithm that tells us how to change integers from one base to another. For example, to change an integer representation in our normal form (base 10) to binary representation (base 2), we repeatedly divide the integer and its successive quotients by 2, keeping track of the remainders. It is these remainders that are the digits of the representation in the new base. For example, since $187 = 93 \cdot 2 + 1$, $93 = 46 \cdot 2 + 1$, $46 = 23 \cdot 2 + 0$, $23 = 11 \cdot 2 + 1$, $11 = 5 \cdot 2 + 1$, $5 = 2 \cdot 2 + 1$, $2 = 1 \cdot 2 + 0$, and $1 = 0 \cdot 2 + 1$, then, $187_{(10)} = 10111011_{(2)}$, which means

$$1 \cdot 10^2 + 8 \cdot 10^1 + 7 \cdot 10^0 =$$
$$1 \cdot 2^7 + 0 \cdot 2^6 + 1 \cdot 2^5 + 1 \cdot 2^4 + 1 \cdot 2^3 + 0 \cdot 2^2 + 1 \cdot 2^1 + 1 \cdot 2^0.$$

The same procedure works for any base b with $b > 1$.
6. We take our lead from the algorithm in the problem immediately above; instead of successively dividing by the same number b, with each division we now increase the divisor by 1. Using the division algorithm, we obtain unique quotients q_i and remainders c_i such that for positive integer m

$$
\begin{aligned}
m &= q_1 2 + c_1 &&\text{with} \quad 0 \le c_1 \le 1, \\
q_1 &= q_2 3 + c_2 &&\text{with} \quad 0 \le c_2 \le 2, \\
q_2 &= q_3 4 + c_3 &&\text{with} \quad 0 \le c_3 \le 3, \\
q_3 &= q_4 5 + c_4 &&\text{with} \quad 0 \le c_4 \le 4,
\end{aligned}
$$

$$\vdots$$

$$
\begin{aligned}
q_{n-2} &= q_{n-1}(n) + c_{n-1} &&\text{with} \quad 0 \le c_{n-1} \le n - 1, \\
q_{n-1} &= q_n(n + 1) + c_n &&\text{with} \quad 0 \le c_n \le n \text{ and } q_n = 0.
\end{aligned}
$$

Since $0 \le q_{i+1} < q_i$, the q_i are decreasing and for some smallest n we must have $q_n = 0$. Iterative substitution will now give us our desired representation, as follows.

$$m = c_1 1! + q_1 2!$$
$$= c_1 1! + (c_2 + q_2 3)2!$$
$$= c_1 1! + c_2 2! + q_2 3!$$
$$= c_1 1! + c_2 2! + (c_3 + q_3 4)3!$$
$$= c_1 1! + c_2 2! + c_3 3! + q_3 4!$$
$$= c_1 1! + c_2 2! + c_3 3! + (c_4 + q_4 5)4!$$
$$= c_1 1! + c_2 2! + c_3 3! + c_4 4! + q_4 5!$$
$$\vdots$$
$$= c_1 1! + c_2 2! + c_3 3! + \cdots + c_n n! + q_n(n+1)!$$
$$= c_1 1! + c_2 2! + c_3 3! + \cdots + c_n n!,$$

where c_j is an integer such that $0 \le c_j \le j$ for $j = 1, 2, \ldots, n$.

§50. Hall's Marriage Theorem.

Homework MI III.
1. Use Hall's Marriage Theorem twice—use the "if part," switch the sexes in the theorem, and then use the "only if part."
2. For $1 \le k \le d$, each k of the destinations (boys) are collectively connected to (collectively know) at least k of the origins (girls).
3. This is a good game of solitaire if you like to be able to win every game. There are 13 columns (boys) and 13 denominations (girls). For example, the 4 cards $5\heartsuit$, $5\spadesuit$, $10\diamondsuit$, $K\clubsuit$ in column i are interpreted as boy i knows the 3 girls in the set $\{5, 10, K\}$. Each collection of k columns has $4k$ cards and so at least k different denominations, since there are only 4 of each denomination. This is the marriage condition and a match is assured.
4. Suppose that each of k boys give us a list of the girls he knows. Counting duplicate names, we have at least kr names. However, since each name is on at most r lists, then we have at least k different names among the k lists. So each set of k boys collectively knows at least k girls. The marriage condition is satisfied and a matching is assured.
5. For each i, replace boy i with g_i different new boys each of whom knows exactly the same girls as boy i and then apply Hall's

theorem. Let A_i be the set of girls boy i knows. The desired condition is that

$$\left| \bigcup_{i \in I} A_i \right| \geq \sum_{i \in I} g_i$$

for all I such that $I \subseteq \{1, 2, \ldots, n\}$.

In their research paper with the elegant proof of Hall's Marriage Theorem, Halmos and Vaughan also solve this problem. They claim this problem is a restatement in the marriage metaphor of "the celebrated problem of the monks." They give Balzac as the source for this "well-known problem." Unable to find the problem in Balzac, this author asked Halmos about the reference. He responded, To the best of my knowledge, Balzac never mentioned any problem of the monks—we invented it—made it up out of whole cloth. Not even the *American Journal of Mathematics* is safe from inside jokes.

6. Suppose at least r of the n boys can be matched with different girls they know. We invent $n - r$ additional girls such that each individually knows all n boys. In this invented situation, there is clearly a matching for the n boys. (The remaining $n - r$ boys can be matched with only the invented girls, for instance.) Conversely, after inventing such $n - r$ girls, if there is a matching in this invented situation, then at least r of the boys must be matched with the original girls since at most $n - r$ of the n boys can be matched with invented girls. Therefore, there is a matching of r boys in the original situation iff there is a matching of all n boys in the invented situation.

Now, by Hall's theorem, there is a matching in the invented situation iff each k of the n boys collectively know at least k girls, both original and invented. Since each boy knows all $n - k$ of the invented girls, the condition is that each k of the n boys must collectively know at least $k - (n - r)$ of the original girls, as desired.

7. The first assumption in the problem is the marriage condition, which assures there is always at least 1 matching. Assuming $r \leq n$, we prove the first conclusion by using the strong form of mathematical induction. The basis step ($n = 1$) is trivial (and points out the limitation $r \leq n$). For the induction step, we assume that the the proposition is true for $n = 1, 2, \ldots, m$ and prove the proposition for $n = m + 1$. We suppose Lucky Pierre is 1 of the

$m + 1$ boys and consider 2 cases. [Case 1.] Assume that for each girl g that Lucky knows there is a matching of the $m + 1$ boys with Lucky paired with girl g and such that the remaining m boys together with all the girls excluding girl g satisfy the marriage condition. So, each of these m boys knows at least $r - 1$ girls other than girl g. By the induction hypothesis, these m boys can be matched in at least $(r - 1)!$ ways. Since this is true for each of the at least r choices for girl g, then there are at least $r(r - 1)!$ matchings for all the $m + 1$ boys in this case, as desired. [Case 2.] Assume that the first case does not hold. Thus, there is a girl, say Lucy, such that if Lucky is paired with Lucy then the marriage condition fails for the remaining boys and girls. This means there must be a set of k other boys that collectively know exactly k girls, namely Lucy and only $k - 1$ other girls, where $k \leq m$. In matching the $m + 1$ boys, these k boys must be matched with these k girls, which by the induction hypothesis can be done in at least $r!$ ways. Hence, there are at least $r!$ ways to match all the $m + 1$ boys with different girls they know. We conclude by the strong form of mathematical induction that the first conclusion in the problem is true for each positive integer n.

We turn to the second conclusion, now assuming $r > n$. It turns out that we will need only ordinary mathematical induction here. We begin by reproducing the proof above, making some necessary changes. The basis step in this situation depends only on the identity $r = \frac{r!}{(r-1)!}$. The corresponding Case 1 now depends on the identity $r\frac{(r-1)!}{((r-1)-m)!} = \frac{r!}{(r-(m+1))!}$. We will be done when we argue that the corresponding Case 2 cannot happen in this situation. Since every boy knows at least r girls, then no set of k boys can collectively know exactly k girls when $k < m + 1 < r$.

Chapter 8. Graphs

§51. The Vocabulary of Graph Theory.

Homework Graphs 1.
1. We would hope that graphs isomorphic to the top 4, having 0, 1, and 2 edges, in Figure 5 would be immediate. Then the middle 3, having 3 edges, would follow with just a little thought. Then, in the best of all possible worlds, we would discover on our own the concept of the complement of a graph. For example, in looking for

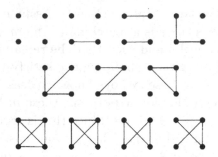

FIGURE 5. The Nonisomorphic Graphs Having 4 Vertices.

selections of 4 of the 6 edges that produce nonisomorphic graphs, we need only select those edges we did not select in forming each of the diagrams in the top row. In general, the **complement** of simple graph G is the simple graph having the same set of vertices as G but such that, for distinct vertices x and y, we have $\{x, y\}$ is an edge of the complement of G iff $\{x, y\}$ is not an edge of G.

2. The vertices of the hexagon can be taken in order as H_1, U_3, H_2, U_1, H_3, and U_2.

3. Consider the simple graph where the people from Kansas are the vertices and 2 vertices are adjacent iff the persons have met. The odd vertices are those persons from Kansas that have met an odd number of persons from Kansas and there must be an even number of them, because there are an even number of odd vertices in every simple graph.

4. Since Q_n has 2^n vertices, since each vertex in Q_n has degree n, and since $2|E_{Q_n}| = \sum_{v \in Q_n} d(v) = 2^n n$, then Q_n has $2^{n-1}n$ edges. For the 3-cube, if we let the 3 vertices adjacent to 000 be in X, the the other 3 vertices must be in Y. No edge of Q_3 has both ends in X or both ends in Y. We have shown that Q_3 is bipartite. Looking at this example, we should guess that, in general, we can let X be the set of vertices of Q_n that have an even number of 0's and Y must then be the set of vertices having odd number of 0's. (We could switch "even" and "odd" here.) No edge can have both ends in X; no edge can have both ends in Y. (Why?) Thus, Q_n is bipartite. (For Q_0, bipartite parts can be given by $X = \{\lambda\}$ and $Y = \emptyset$, where λ is the empty sequence and $\emptyset$ is the empty set.)

5. We are aware that there are $2^{\binom{k}{2}}$ simple graphs having k given vertices, since each of the $\binom{k}{2}$ possible pairs of vertices either is or

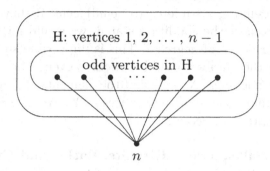

$$d(n) = \# \text{ odd vertices in } H.$$

FIGURE 6. Graphs with all Vertices Even.

is not an edge of the graph. First, given any 1 of the $2^{\binom{n-1}{2}}$ simple graphs H having $\{1, 2, 3, \dots, n-1\}$ as its set of vertices, we form a new graph G by adding n as the new vertex and by adding all edges such that n is now adjacent to all the odd vertices in H. So, all the odd vertices in H become even in G. Beware, here as in Figure 6, "even" and "odd" refer to the degree of a vertex and not to the vertex itself. (For example, vertex 2 may be an odd vertex in H.) The even vertices in H are still even in G. The new vertex n is also even in G because the number of odd vertices in H is necessarily even. (The number of odd vertices in any simple graph is even.) So all the vertices of the new graph G are even, as desired. (In symbols, this is $V_G = V_H \cup \{n\}$ and $E_G = E_H \cup \{\, \{n, v_j\} \mid v_j \text{ is odd in } H \,\}$.) Conversely, given a graph G having $\{1, 2, 3, \dots, n\}$ as its set of vertices and such that each vertex has even degree, we can easily form a simple graph H having $\{1, 2, 3, \dots, n-1\}$ as its set of vertices by deleting vertex n and all edges adjacent to n from G. Thus, there is a one-to-one correspondence between the set of simple graphs having $\{1, 2, 3, \dots, n\}$ as its set of vertices and such that each vertex has even degree and the set of simple graphs having $\{1, 2, 3, \dots, n-1\}$ as its set of vertices.

6. The graph *is* $K_{3,3}$ and, so, is not planar. Begin by labeling, in order, the vertices of the square as H_1, U_1, H_2, and U_2. This determines H_3 and U_3.

7. Consider the simple graph formed by taking the 7 students as vertices and defining 2 students to be adjacent iff they exchanged postcards. To meet the condition stated, we would have a graph with 7 vertices, each of degree 3. This is impossible since in every simple graph the number of odd vertices is even. The answer is 0.

8. Call the elements of X boys and the elements of Y girls. Let r be the maximum degree of the vertices in Y. We now have problem #4 from the previous section.

§52. Walks, Trails, Circuits, Paths, and Cycles.

Homework Graphs 2.

3. There are lots of possibilities. For example: (a) ⊔, (b) ⋈, (c) ⊔⊔, (d) ⊔.

4. Start drawing from a vertex and, without lifting the pencil from the paper, wander around the page, passing through previous intersections at will, and stop at some other vertex. Now, mark as vertices the intersections and some additional points, as necessary, to eliminate parallel edges and loops.

5. ⊔⊔ and |⊏⊐| are hamilton paths. Trying to construct a hamilton cycle, we begin by noticing that the 4 corner vertices and their adjacent edges must be part of any hamilton cycle. (Each vertex of degree 2 must be part of every hamilton cycle.) Now, of the remaining 4 edges around the outside of the figure, not all 4 can be part of the hamilton cycle. Picking any of the alternative edges determines edges until an impasse is reached.

7. Since each vertex in Q_n has degree n, then Q_n has an euler circuit iff $n \geq 2$ and n is even.

9. See Figure 7.

10. Our proof that an n-cube has a hamilton cycle is by mathematical induction on n. Follow the argument with a 3-cube (an ordinary cube) and with the solution in Figure 7 for the representation of a 4-cube (a tesseract) given in question #9. The base step $n = 2$ is trivial, since Q_2 *is* a cycle. For the induction step, we assume that a k-cube has a hamilton cycle for some k such that $k \geq 2$ and prove that a $(k+1)$-cube has a hamilton cycle. Given the $(k+1)$-cube, consider the k-cube K formed by all vertices ending in 0 and the edges of the $(k+1)$-cube that join 2 such vertices. For each such vertex v, let v' denote the vertex differing only in the last digit. So, v' ends in 1 and vertices v and v'

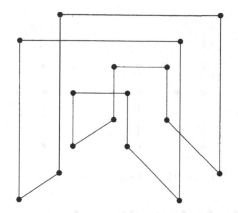

FIGURE 7. A Hamilton Cycle on the Tesseract.

are adjacent. With $K' = \{v' \mid v \in K\}$, the set K' together with the edges of the $(k+1)$-cube that join these vertices also forms a k-cube. By the induction hypothesis, there are hamilton cycles

$$a\{a,b\}b\{b,c\}c\cdots z\{z,a\}a \quad \text{and} \quad a'\{a',b'\}b'\{b',c'\}c'\cdots z'\{z',a'\}a'$$

in K and K', respectively. We replace edges $\{a,b\}$ and $\{a',b'\}$ by $\{a,a'\}$ and $\{b',b\}$ and arrange the terms as.

$$b\{b,c\}c\cdots z\{z,a\}a\{a,a'\}a'\{a',z\}z'\cdots c'\{c',b'\}b'\{b',b\}b,$$

which is a hamilton cycle in the $(k+1)$-cube. The result follows by mathematical induction.

§53. Trees.

Homework Graphs 3.
1. See Figure 8.

2.

There are only 6 candidates, given in Figure 8. All trees are bipartite. However, if we start labeling the vertices alternately R (for red) and G (for green), then we must end with 3 R's and 3 G's to have a spanning tree for $K_{3,3}$. This eliminates the second, fifth, and sixth of the trees in Figure 8, leaving the 3 possibilities.

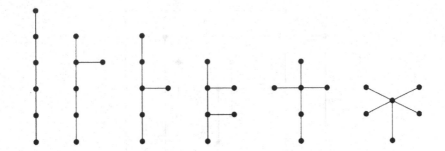

FIGURE 8. Trees with 6 Vertices.

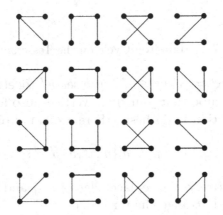

FIGURE 9. Spanning Trees for K_4.

3. Since there are 14 leaves, the vertices of degree 4 must be connected by $(6 \cdot 24 - 14)/2$ edges. So, the 6 (carbon) vertices of degree 4 must be connected by $6 - 1$ edges and thus must themselves form a tree, to which the 14 (hydrogen) leaves can be attached. We return again to Figure 8. Since there are no vertices of degree 5, the last tree in the figure must be eliminated. However, each of the first five trees is is a feasible tree for the the vertices of degree 4. There are 5 solutions.

4. By Cayley's Theorem, we expect 4^2 spanning trees. See Figure 9.

5. 3333999,   2112112,   2,6,2,2,4,6,6,6,6,5,4,4,3,2,2.

6.

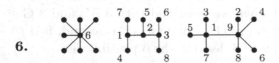

7. Suppose that $\{a, b\}$ and $\{v_1, v_2, \ldots, v_n\}$ are bipartite parts for $K_{2,n}$. In a spanning tree, there is a unique path from a to b, and since $\{a, b\}$ has only 2 elements, then this path must be of length 2. There are n choices for this path $a\{a, v_i\}v_i\{v_i, b\}b$. Then the remaining $n - 1$ vertices must be connected to exactly 1 of a or b in order to have a tree that spans. These choices can be made in 2^{n-1} ways. Hence, altogether, we have a total of $n2^{n-1}$ possible spanning trees.

§54. Degree Sequences.

Homework Graphs 4.

1. With n vertices the possible degrees are the n possible values $0, 1, 2, \ldots, n - 2, n - 1$, but 0 and $n - 1$ are impossible together. We have n vertex pigeons and $n - 1$ pigeonholes.

2. Neither 42200 nor 32100 is graphic.

3. The graphic sequences in reverse order are 1100, 42211, 553322, and 6664433.

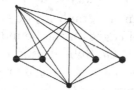

4. Sequence 555553 is only 1 possibility.

5. Here, we let H_k denote a graph with n vertices, each of degree 4. Clearly, $H_5 = K_5$. The existence of H_6, H_7, H_8, and H_9 follows from the degree sequence theorem. The rest follow by mathematical induction. For $k > 9$, we take the union of the graphs H_5 and H_{k-4}. For a different attack, start with C_n (an n-gon) and join alternate vertices. Further, when n is even, say $n = 2k$, we can get a planar graph by joining each vertex of a regular k-gon to the closest 2 vertices of a smaller, concentric k-gon rotated $(180/k)°$. For the case $k = 4$, we have the following graph.

6. An n-gon C_n will do nicely.

7. The number of vertices must be even since each vertex has odd degree. Here, we take a $2k$-gon and add the k diagonals. Actually, we can do better and find planar graphs. Graph K_3 is a special case $k = 2$, with 1 vertex in the interior of the triangle formed by the other 3 vertices. For $k > 2$, we can take a regular k-gon with a concentric larger k-gon and join corresponding vertices. For example, for the case $k = 4$ we have the following graph.

§55. Euler's Formula.

Homework Graphs 5.

1. Since there are neither loops nor parallel edges in a planar graph, every region in a diagram must be bounded by at least 3 edges. Each edge is the boundary of at most 2 regions; some edges such as the right edge in the graph ⊿ bound only 1 region. So, summing the number of edges on the boundary for each of the r faces, we must have $3r \leq 2q$. Substituting $r \leq 2q/3$ in Euler's Formula, we get $2 = p - q + r \leq p - q/3$ or $q \leq 3p - 6$.

2. Assuming K_5 is planar gives the following contradiction to Corollary 1. $10 = \binom{5}{2} = q \leq 3p - 6 = 3 \cdot 5 - 6 = 9$.

3. If our planar graph G has no triangles, meaning that K_3 is not a subgraph of G, then each region in a diagram must have at least 4 edges as boundaries. Summing the number of edges on the boundary for each of the r faces, this time we get $4r \leq 2q$. Substituting $r \leq q/2$ in Euler's Formula, we get $2 = p - q + r \leq p - q/2$ or $q \leq 2p - 4$.

4. Since $K_{3,3}$ is bipartite and so has no triangles, then assuming $K_{3,3}$ is planar gives the following contradiction to Corollary 3. $9 = 3 \cdot 3 = q \leq 2p - 4 = 2 \cdot 6 - 4 = 8$.

5. Suppose our graph G has no vertex of degree less than 6. Summing the degree of the vertex for each of the p vertices, we get $6p \leq 2q$. Hence, by Corollary 1, we then have the contradiction $3p \leq q \leq 3p - 6$. Therefore, there must be at least 1 vertex having degree less than 6.

Index

Undergraduate Texts in Mathematics

(continued from page ii)

Halmos: Naive Set Theory.
Hämmerlin/Hoffmann: Numerical
    Mathematics.
    Readings in Mathematics.
Harris/Hirst/Mossinghoff:
    Combinatorics and Graph Theory.
Hartshorne: Geometry: Euclid and
    Beyond.
Hijab: Introduction to Calculus and
    Classical Analysis.
Hilton/Holton/Pedersen: Mathematical
    Reflections: In a Room with Many
    Mirrors.
Hilton/Holton/Pedersen: Mathematical
    Vistas: From a Room with Many
    Windows.
Iooss/Joseph: Elementary Stability
    and Bifurcation Theory. Second
    edition.
Isaac: The Pleasures of Probability.
    Readings in Mathematics.
James: Topological and Uniform
    Spaces.
Jänich: Linear Algebra.
Jänich: Topology.
Jänich: Vector Analysis.
Kemeny/Snell: Finite Markov Chains.
Kinsey: Topology of Surfaces.
Klambauer: Aspects of Calculus.
Lang: A First Course in Calculus. Fifth
    edition.
Lang: Calculus of Several Variables.
    Third edition.
Lang: Introduction to Linear Algebra.
    Second edition.
Lang: Linear Algebra. Third edition.
Lang: Undergraduate Algebra. Second
    edition.
Lang: Undergraduate Analysis.
Lax/Burstein/Lax: Calculus with
    Applications and Computing.
    Volume 1.
LeCuyer: College Mathematics with
    APL.
Lidl/Pilz: Applied Abstract Algebra.
    Second edition.

Logan: Applied Partial Differential
    Equations.
Macki-Strauss: Introduction to Optimal
    Control Theory.
Malitz: Introduction to Mathematical
    Logic.
Marsden/Weinstein: Calculus I, II, III.
    Second edition.
Martin: Counting: The Art of
    Enumerative Combinatorics.
Martin: The Foundations of Geometry
    and the Non-Euclidean Plane.
Martin: Geometric Constructions.
Martin: Transformation Geometry: An
    Introduction to Symmetry.
Millman/Parker: Geometry: A Metric
    Approach with Models. Second
    edition.
Moschovakis: Notes on Set Theory.
Owen: A First Course in the
    Mathematical Foundations of
    Thermodynamics.
Palka: An Introduction to Complex
    Function Theory.
Pedrick: A First Course in Analysis.
Peressini/Sullivan/Uhl: The Mathematics
    of Nonlinear Programming.
Prenowitz/Jantosciak: Join Geometries.
Priestley: Calculus: A Liberal Art.
    Second edition.
Protter/Morrey: A First Course in Real
    Analysis. Second edition.
Protter/Morrey: Intermediate Calculus.
    Second edition.
Roman: An Introduction to Coding and
    Information Theory.
Ross: Elementary Analysis: The Theory
    of Calculus.
Samuel: Projective Geometry.
    Readings in Mathematics.
Scharlau/Opolka: From Fermat to
    Minkowski.
Schiff: The Laplace Transform: Theory
    and Applications.